META-X®-Software for Metapopulation Viability Analysis

UFZ – Centre for Environmental Research Leipzig-Halle (Ed.)

Karin Frank Helmut Lorek Frank Köster Michael Sonnenschein
Christian Wissel Volker Grimm

Springer-Verlag Berlin Heidelberg GmbH

UFZ – Centre for Environmental Research Leipzig-Halle (Ed.)
Karin Frank   Helmut Lorek   Frank Köster
Michael Sonnenschein   Christian Wissel   Volker Grimm

META-X®-Software for Metapopulation Viability Analysis

Springer

Dr. Karin Frank
Professor Dr. Christian Wissel
Dr. Volker Grimm
UFZ – Centre for Environmental Research Leipzig-Halle
Dept. of Ecological Modelling
P.O. Box 500136
04301 Leipzig
Germany

Dr. Frank Köster
Carl von Ossietzky Universität Oldenburg
Fachbereich Informatik
Escherweg 2
26121 Oldenburg
Germany

Dr. Helmut Lorek
Math Consulting Group AG
Bahnhofsstrasse 17
6301 Zug
Switzerland

Professor Dr. Michael Sonnenschein
OFFIS – Oldenburg Research and Development Institute
for Computer Science Tools and Systems
Escherweg 2
26121 Oldenburg
Germany

Additional material to this book can be downloaded from http:// extras .springer .com .

ISBN 978-3-642-62906-8   ISBN 978-3-642-55723-1 (eBook)
DOI 10.1007/978-3-642-55723-1

Library of Congress Cataloging-in-Publication Data applied for

A catalog record for this book is available from the Library of Congress.

Bibliographic information published by Die Deutsche Bibliothek
Die Deutsche Bibliothek lists this publication in the Deutsche Nationalbibliographie;
detailed bibliographic data is available in the Internet at http://dnb.ddb.de

http://www.springer.de

© Springer-Verlag Berlin Heidelberg 2003
Originally published by Springer-Verlag Berlin Heidelberg New York in 2003
Softcover reprint of the hardcover 1st edition 2003

Cover design: *Erich Kirchner*, Heidelberg
Typesetting: Camera-ready by the editors

31/3150 YK – 5 4 3 2 1 0 – Printed on acid free paper

Preface

The user-friendly META-X software for metapopulation viability analysis provided with this handbook is the result of fruitful cooperation between the Department of Ecological Modelling of the UFZ Centre for Environmental Research Leipzig-Halle and the OFFIS Oldenburg Research Institute for the Development of Computer Tools and Systems. The Department of Ecological Modelling has long experience of the development and analysis of standard models for population viability analysis. Therefore, it made sense to seek out opportunities to make this knowledge available and applicable for teaching, research as well as conservation practice and landscape management. At the same time, in 1996, the OFFIS-Institute was looking for a partner with experience in ecology to help develop computer tools for environmental management. This was the beginning of a five-year cooperation project mainly funded by the UFZ (grant code UFZ-21/96). The idea was to use the generic metapopulation model presented in Chapter 14 and the standard procedure for determining population viability described in Chapter 13 as the heart of the software and to develop a user interface geared to the workflow of decision-making in the field of nature conservation and landscape management. The aim was to come up with a tool that supports understanding, ranking management options and dealing with ecological uncertainty.

Throughout this time, the META-X team received much assistance from several people who made essential contributions to the project's success in several respects. First of all, we would like to thank Michael Malachinski and Jens Finke who programmed some key features of META-X such as the landscape editor and the report generator. We are also grateful to Atte Moilanen from the University of Helsinki who placed his method of parameterizing metapopulation models on the basis of occupancy data at our disposal and adapted his procedure to the specific requirements of META-X. This procedure provides the basis for the parameterization tool which can be downloaded from the META-X homepage. All the programming work on the corresponding user interface was done by Torsten Abels to whom we express our gratitude. Many thanks also go to Ilse Storch who provided the data, the expert knowledge and the questions to be answered by META-X regarding the example of the capercaillie presented in Chapter 16.

In the second half of the project, the various versions of META-X were extensively tested for correctness by comparing the results of the software with those of the original model which was implemented as a *Mathematica* program. We are most grateful to Alexander Singer for conducting this long-term test with enormous patience and precision. He also implemented and analyzed all the examples presented in this handbook. We would also like to thank Simone K. Heinz who particularly tested the import routine of META-X. Many thanks to Ute Vogel from the University of Oldenburg for fruitful discussions on the design of the user interface of Meta-X and to Monika Schwager and Matthias Wichmann from the University of Potsdam who tested the software from the user's point of view and who also commented on earlier drafts of the entire handbook. Moreover, we would like to thank Jürgen Groeneveld, Simone K. Heinz, Stephanie Schadt and Björn Reineking for carefully reading and critically commenting on Chapter 13 of this book.

Last but not least, we are grateful to Chris Abbey for polishing the English and Ursula Gramm from Springer Verlag for her vital hints concerning the handbook's layout.

We hope very much that you are satisfied with the results – and that META-X lives up to your expectations!

The META-X team

Leipzig and Oldenburg, July 2002

Contents

1 Introduction

Goals

This introduction answers the following questions:

- What is META-X?
- What is so special about META-X?
- In what areas can META-X be used?
- For whom has META-X been designed?
- What do you need to use META-X?
- What cannot be done with META-X?

1.1 What Is META-X?

A metapopulation is defined as a 'population of populations' which go extinct locally and recolonize. META-X is a program which helps assess the probability of survival or the risk of extinction of metapopulations. The program implements a simple, generic model of a metapopulation. By choosing appropriate parameters, this model can be tailored to certain general questions or to real metapopulations in actual landscapes. Once this has been done, META-X comfortably supports all the steps of model-based population viability analysis (PVA).

The three basic terms set in bold type in this short description of META-X are explained in more detail below.

Metapopulation

A metapopulation is a group of two or more local populations or subpopulations which live on discrete habitat islands (patches).

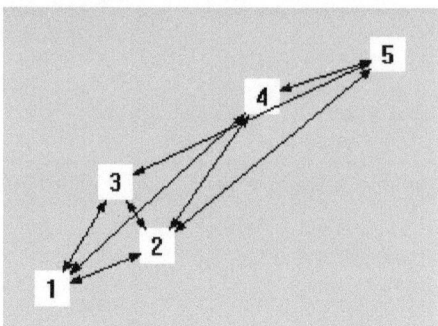

Setup of a metapopulation

Although this fragmentation of habitat into discrete patches may be natural, nowadays it is caused for most species by human impacts. Very often, the subpopulations are not large enough to persist for a longer time, i.e. they are not viable. Sooner or later they will go extinct due to random fluctuations in environmental conditions and demographic processes (birth and death).

After a subpopulation has gone extinct due to random fluctuations, its habitat will still exist and may be recolonized by individuals leaving other patches which are still occupied by subpopulations. If the rate of recolonization is sufficiently larger than that of local extinctions, the whole metapopulation may persist much longer than even its most long-lived subpopulation.

PVA

Population viability analysis (PVA) targets three goals, which are closely interrelated:

1. Identifying and analyzing the processes and structures of (meta)populations which determine their risk of extinction. How do these processes and structures interact, and which process and which structural element of the landscape exerts the strongest threat to the metapopulation?
2. Quantifying extinction risk – because conservation and environmental management need 'safety in numbers' if they are to withstand other societal forces which may lead to the extinction of populations and species. Vague allusions to 'too small' habitats and 'too high' extinction risks are inadequate.
3. Management decisions which change the extinction risk of population on a rational basis. To this end, the extinction risks are quantified for all alternative decisions and ranked: which decision is likely to benefit the population most, and which decision will probably bring about the greatest threat to the population?

The main tool used by PVA to achieve these goals is ecological models.

Ecological Models

Ecological models are purposeful representations of problems or questions. They should not be interpreted as realistic images of nature but rather as tools for problem-solving.

Models emerge as soon as we consider a problem: certain aspects of reality are taken into account because they are considered essential, whereas other aspects are neglected since they are considered less important. Very often, models are used if information required to solve the problem is unavailable but the problem still has to be solved. Gaps in knowledge are taken into account by filling them with different assumptions and then exploring the logical consequences of these assumptions.

Since verbal and graphical models are hard to test for completeness and consistency, the models applied in PVA use the formal language of mathematics. For problems with a very simple structure, mathematical methods can be used to calculate results (e.g. to assess extinction risks) from the model. More frequently, however, computers are used. The ecological model is – as is also the case with META-X – translated into a computer program.

1.2 What Is so Special About META-X?

The special features of the model underlying META-X are:

- META-X concentrates on regional processes (recolonization, spatially correlated extinction of subpopulations) and therefore describes the local dynamics of the subpopulations in a highly aggregated way. This keeps the number of model parameters small.
- Spatial correlations of local extinctions can be taken into account.
- Peculiarities of the landscape, such as barriers to dispersal and corridors, and structures which affect spatial correlation, can be taken into consideration.

The special features of the concept behind META-X are:

- It supports the definition, handling and joint evaluation of comparative computer experiments. PVA does not aim to produce absolute assessments of extinction risks, i.e. individual numbers, but rather relative assessments. It is precisely this kind of relative, comparative assessment which is implemented in META-X.

The special features of the program META-X are:

- The input of model parameters is facilitated by wizards, i.e. step-by-step instructions on how to proceed. Parameters of external submodels may be imported.
- META-X allows the automatic variation of model parameters.

- A graphical Landscape Editor visualizes the landscape specified by the model parameters. The landscape editor allows the number and position of patches, the connections between patches and the properties of the subpopulations to be newly defined or modified.

The special feature of the results produced by META-X is:

- The use of the 'intrinsic mean time to extinction' as a basic measure to quantify persistence which can then be employed to calculate extinction risk for any time horizon. Other output quantities characterize the importance of each patch and the ability of the whole metapopulation to recover.

1.3 In What Areas Can META-X Be Used?

In Teaching

META-X is an ideal tool to teach and learn almost all aspects of metapopulation theory. A thorough understanding of this theory is now a must for ecologists and in particular for conservation biologists and planners. Teachers can use META-X to compile a course about PVA and metapopulations. Students can use META-X to practice or teach themselves PVA and metapopulation theory. The advantage of META-X for teaching and learning is that no skills in mathematics or programming are required to use META-X successfully.

In Empirical Research

Empirical studies of metapopulations are often saddled with the problem that conclusions about long-term dynamics are hard to achieve from short-term field studies or snapshots. Although under optimal circumstances, a tailored ecological model could be developed to solve this problem, there are usually insufficient resources (time, money, qualified personnel) to do so. META-X, however, can be used by empirical researchers by themselves. META-X allows hypotheses about the spatiotemporal dynamics of a metapopulation to be explored, along with the relative importance of different processes and structures in a landscape. This enables the limited information available to be integrated and extrapolated to achieve the best possible understanding of the long-term dynamics of the metapopulation. In turn, integrating empirical information and understanding will help subsequent empirical studies to be designed in a more focused, efficient manner.

In Biological Conservation and Planning

However, the main fields of application of META-X are biological conservation and planning. In addition to the lack of resources for developing tailored models and PVA being at least as high as in research, there are two more problems:

1. In most cases, even less information exists about the population in question than in empirical research.
2. Since PVA and metapopulation theory are relatively new approaches, there is still a lack of basic understanding of these approaches among conservation biologists and planners. So far, this lack of understanding has prevented the broader and more sophisticated application of these approaches.

META-X takes both these problems into account. Because of the first problem, META-X concentrates on regional aspects of metapopulation dynamics and thereby keeps the number of model parameters manageable. By using, for example, expert assessments of these parameters, complete PVAs can be performed. The uncertainty of expert assessment is taken into account by a concept which is basic to META-X, namely comparative experiments, i.e. META-X is designed to optimally support sensitivity analyses, the automatic variation of individual parameters and the comparison of scenarios. As an alternative to expert assessments, external submodels can be used to determine model parameters. META-X allows these parameters to be imported and is thus open to all kinds of external models and tools.

The second problem is solved by META-X providing (see above: 'In teaching') the opportunity to study PVA and metapopulations by oneself.

1.4 What Do You Need to Use META-X?

Successfully using META-X entails the following requirements:

- The hardware requirements are a personal computer with a Pentium Processor (or 100% compatible) with at least 133 MHz, 64 MB RAM, a hard disk with at least 15 MB free space, and a CD-ROM drive with at least 4x speed. The software requirements are: Microsoft Windows 95/98, Windows 2000, Windows NT 4.0 or Windows XP, and Microsoft Internet Explorer 4.0 or higher.
- You should be familiar with programs running on Windows 95/98/2000/NT/XP (windows, mouse-clicks, icons, dialog boxes, etc.). If this is not the case, this can all be learned in under half an hour by following the tutorial included with the various Windows operating systems.
- You should be familiar with the basic concepts, goals and methods of PVA. Only those who know and understand the goals and methods of PVA will be able to perform PVA and to benefit from META-X. If you do not have this knowledge, please read the Chapter 13 and, if necessary, the other literature

cited in that chapter. After reading this, you will be able to explore the most important potentials and limitations of PVA yourself by using META-X.

- Make sure you read Chapters 4 and 5 ('Basic Concepts' and 'Guided Tour') and follow the Guided Tour by using META-X.
- 'Play around' with META-X as much as possible in order to get a feel for the program! As an explorative, creative beginner you will rapidly learn what can be done with META-X – and how to do it. The Chap. 16 ('Example Applications') also gives ideas for exploring META-X.

There are no other requirements for using META-X. In particular, working with META-X does not require special skills in mathematics or programming.

1.5 What Cannot Be Done with META-X?

Tailored ecological models which are developed for specific landscapes and specific metapopulations have the advantage that they can take into account all the processes and structures which are considered essential in this specific case. There are, however, many more peculiarities and special cases than can be taken into account in a generic, broad-purpose tool like META-X. The model underlying META-X is a compromise between applicability to as many cases as possible and the potential to allow peculiarities to be included.

Thus, PVAs based on META-X will support decision-making in conservation and planning with less weight than PVAs based on tailored models. META-X cannot replace the decision if a tailored model is to be developed or if no PVA is to be performed at all or if META-X should be used as an intermediate alternative between these two extremes. Nevertheless, even this decision could be guided by META-X, because META-X gives swift insights into the general potentials and limitations of PVA, as well as into the amount of empirical information required before META-X or any other model could be used for PVA.

It should, however, be emphasized that without sound assessments of the majority of model parameters of META-X, reliable assessments of extinction risks cannot even be produced with META-X. The Chapter 15 ('Parameterizing META-X') explains in detail the ways to parameterize META-X.

Finally, it should be noted that META-X – like any PVA – is only a tool for decision-making in conservation and planning. The tool itself does not make the decisions – this remains the responsibility of humans.

1.6 News on META-X

We are planning to maintain a META-X webpage as a forum for META-X users to exchange ideas, submodels, example applications etc. We will provide special tools to parameterize META-X (for example to derive parameters from presence-absence data). Likewise, users are welcome to send us their own tools and submodels (or links to these tools) to be posted on the page for downloading. We thus hope that the webpage will become a forum that helps integrate the activities of different users of META-X and other models, programs and tools for PVA. The URL of the META-X webpage is: http://www.oesa.ufz.de/meta-x

2 Contents

Goals

The META-X book consists of two parts which are easily distinguished by their differing layout:

Part I is the META-X manual, which provides an introduction to using the program META-X. Its layout is marked by numerous screen shots. It provides:

- A tutorial for the usage of META-X.
- An introduction to advanced features of META-X in separate chapters.
- A reference to all the features of META-X.

Part II introduces basic concepts and methods of PVA, the model which is implemented in META-X, and applications using META-X. The contents of the chapters in the META-X book are briefly described below.

2.1 Part I

Installation: This part describes the installation of META-X from the CD-ROM to your computer, along with how to deinstall META-X.

Basic Concepts: This chapter introduces the basic concepts of META-X: the model underlying META-X, organizing your work with META-X in projects, (computer) experiments and scenarios and the steps to be performed when using META-X.

Guided Tour: This step-by-step tutorial takes you through the most important elements of META-X. Once you have completed the Guided Tour and have read Chapter 4, you will be able to explore the more advanced features of META-X on your own. The other chapters of the META-X book will assist you in this.

Project Tree: Each project is in fact a 'Project Tree' – a hierarchical structure consisting of experiments, scenarios, parameters and evaluations. The Project Tree is visualized in the META-X window and can be used for navigating through projects and for organizing and modifying them.

Scenarios and Experiments: Here, the two basic units of META-X, scenarios and experiments, are described in detail. The focus is on inputting model parameters, and on the definition, modification and management of experiments.

Simulation and Evaluation: 'Simulation' means running the computer for one or more sets of model parameters. With META-X, simulations may be run interactively or automatically. This chapter explains how this can be done. The results of the simulations, i.e. the results of the computer experiment, are then summarized and presented by META-X in 'evaluations'. This chapter explains how evaluations are to be interpreted and used for your further work.

Landscape Editor: The Landscape Editor is an additional graphical tool for defining, modifying and visualizing landscapes which consist of patches (= habitat islands) and connections between these patches.

Parameter Variation: In global and local sensitivity analyses, only one single model parameter is changed while the other parameters are kept constant. META-X supports the automatic definition of variation experiments, i.e. the automatic variation of individual model parameters.

Import, Export and Report: META-X automatically compiles reports of your (computer) experiments and evaluations. These reports are in HTML format.

Reference: This chapter lists the menu of META-X, its context menus and the main warning messages.

2.2 Part II

Goals, Methods and Concepts of PVA: If you are not familiar with the goals, methods and concepts of population viability analysis (PVA), you should read this chapter before you start using META-X. This chapter also contains annotated references to more advanced scientific literature about PVA.

The META-X Model in Detail: Using META-X means working with a generic model of metapopulations. This model is described and explained in detail, including all the mathematical formulas and the computer algorithms used.

Parameterizing META-X: The most difficult and time-consuming task in applying META-X to real metapopulation and landscapes is specifying the model parameters. Although for the purposes of familiarizing yourself with the program and teaching it is sufficient to simply assume parameters, for applications in conservation and planning, parameters must be determined which are as reliable as possible. It is explained how these parameter estimates can be obtained from the information available, possibly by using simple submodels.

Example Applications: To illustrate what can be done with META-X, and how, example applications are given considering extinction theory, planning and the viability assessment of specific populations.

Glossary: The glossary gives concise explanations of the numerous terms used in PVA and META-X which might be unfamiliar to you.

Index: The two-level index helps you to navigate through the META-X book.

Online Help: Part I of the META-X book and the online help of META-X are identical – with one exception: the Guided Tour is only in the book, not the online help.

3 Installation

Goals

The installation and deinstallation of META-X are described. META-X takes advantage of the Windows 32-bit architecture. It is not compatible with earlier versions of Windows, such as Windows 3.11.

3.1 System Requirements

META-X requires

- A personal computer compatible with Windows 95/98/98SE or Windows NT4.0/2000 or Windows XP.
- A Pentium processor (or compatible) with at least 133 MHz clock frequency.
- At least 64 MByte of RAM memory.
- 15 MByte of free disc space.
- A CD-ROM drive.
- A Microsoft Windows-supported mouse and monitor.
- Microsoft Internet Explorer 4.0 or higher for displaying the online help of META-X and for displaying the reports generated by META-X.

3.2 Installation

When you are ready to install META-X, follow these steps:

1. Start Windows.
2. Insert the META-X CD into your CD-ROM drive. The installation procedure may run automatically at this time.
3. If the installation procedure does not run automatically, open the Windows Explorer, and then change to the CD-ROM directory and double click on setup.exe.

4. You are guided by a wizard through the installation procedure. The installation of the software on your computer requires accepting the META-X licence agreement.
5. The wizard allows you to install META-X in a directory of your choice. You can define a new program group for META-X or use an existing program group. The installation does not create any icon on the desktop of your computer.
6. At the end of the installation you may start META-X immediately or return to Windows.

Note that in this book, the buttons controlling the program – e.g. **Next**, **Cancel**, **Yes**, **No** – are presented in English but may be different on your computer owing to another language chosen to be used for Windows.

3.3 Deinstallation

You may remove META-X by the appropriate routines of Windows:

1. Select the software folder in the control panel of your computer by a double click.
2. Select META-X in the list of installed software.
3. Choose the remove option by a click on the appropriate button.
4. META-X and all of its components will be removed from your computer.
5. Close the software window.

4 Basic Concepts

Goals

Before you start working with META-X or reading the following chapters, you should take the time to read this chapter. It introduces the basic concepts and steps of META-X and thus forms the basis for the rest of this manual and for all work with META-X.

4.1 The Model

META-X is designed to be employed by as many different users as possible for as many different problems as possible in teaching, research, conservation and planning. Therefore, the structure of the model underlying META-X is very general. The model considers only those elements which are known – from theory – to be essential to assess the persistence of metapopulations. Consequently, the META-X model is not complex and has a clear structure.

These elements can be grouped into three blocks:

1. **Landscape structure**, i.e. the number and arrangement of patches (habitat islands) in the landscape, possible barriers to dispersal and other peculiarities of the landscape.
2. **Patch characteristics**, i.e. features which characterize the patches and the subpopulations living on the patches.
3. **Spatial processes**, i.e. the dispersal of individuals which may lead to the recolonization of unoccupied patches, and the correlation of extinction events on patches more or less closely located to each other.

4.2 Landscape Structure

In META-X, a landscape consists of two or more patches on which subpopulations of the investigated species may become established, as well as of connections between these patches.

Position of Patches

The location of a patch is specified by its X and Y coordinates in a coordinate system, which is defined by the user and which represents a clipping of the landscape of, say, 100×100 km^2. The patches are always visualized as circles with the X and Y coordinates of the centre of the circles defining the position of the patches.

Initial Occupancy

For each patch it is specified whether at the beginning of a model run (i.e. simulation; see above) the patch is occupied by a subpopulation.

Net of Connections

For metapopulation dynamics, the question of from which occupied patch an unoccupied patch may be recolonized is decisive. Dispersal barriers may exist which prevent recolonization despite comparatively low patch-to-patch distances. To take this into account in META-X, the net of connections has to be explicitly defined, i.e. for each pair of patches it has to be specified whether recolonization is, in principle, possible there.

Note: Other peculiarities of the landscape which may modify the mutual colonization rate of pairs of patches can be taken into account in the reachability matrix and correlation matrix (see Chap. 7).

4.3 Patch Characteristics

Patches are not characterized by their shape or size, but only by the three main aspects which are decisive for the persistence of metapopulations:

Local Extinction Rate

The local extinction rate of a certain patch gives the risk of extinction per year for a subpopulation inhabiting a certain patch. Large patches or patches with optimal habitat quality usually allow for larger subpopulations which, in turn, generally have a smaller risk of extinction (Chap. 13). A quantity being inverse to the rate of extinction is the 'intrinsic mean time to extinction'. If, for example, the intrinsic mean time to extinction is 100 years, i.e. the subpopulation survives on average for 100 years, the risk of extinction per year is approximately $1/100=0.01$.

   The META-X model uses the parameter 'local extinction rate' to determine stochastically for each patch in every year whether the subpopulation goes extinct.

   There are different ways to assess this and the other model parameters (Chap. 15):

- By using a submodel for a spatially unstructured population.
- By using values published in the literature.
- Using assessments by experts who know the species in question well.
- By using both optimistic and pessimistic estimates of the parameter and comparing the outcome of the different results. META-X supports this kind of comparative analysis.

For teaching purposes and for beginners who want to familiarize themselves with META-X, it is sufficient simply to assume local extinction rate by, for example, contrasting important patches characterized by a low extinction rate with (seemingly) less important patches and thus less 'safe' patches.

Number of Emigrants

This parameter specifies the mean number of emigrants an occupied patch produces per year, i.e. individuals who leave their home territory to (possibly) colonize other, empty patches.

Number of Immigrants Needed

Theory and experience show that a certain minimum number of individuals is required to establish a new subpopulation on an unoccupied patch. The parameter 'number of immigrants' specifies how many individuals are needed to establish a subpopulation with a probability of 50%. Therefore, the final recolonization rate of an empty patch results from the product

$$0.5 \cdot \frac{\text{number of immigrants arriving per time}}{\text{number of immigrants needed}}.$$

4.4 Spatial Processes

The model underlying META-X takes into account the two spatial processes which – besides the patch characteristics – are decisive for metapopulation persistence:

Dispersal

The potential for the dispersal of emigrants is characterized by the mean dispersal distance. This distance is converted to 'reachability', i.e. the probability that an emigrant starting from a certain source patch reaches a certain target patch. This probability decreases with increasing distance between two patches.

The model uses reachability, the net of connections, the number of emigrants and the number of immigrants to calculate the main model parameter, the rate of colonization, for each pair of patches. The resulting 'colonization matrix' is displayed to the user and may – like other parameter matrices as well – be modified manually to take into account peculiarities of the landscape:

	P1	P2	P3	P4	P5
P1		0.03115	0.03706	0.01589	0.00000
P2	0.01804		0.04396	0.02523	0.00541
P3	0.01417	0.02903		0.00000	0.00441
P4	0.07943	0.21789	0.00000		0.14032
P5	0.00000	0.03436	0.04245	0.10318	

Correlated Extinctions

An ensemble of interconnected patches, i.e. a metapopulation, may persist for a much longer time than even its most long-lived single subpopulation. This is because the extinction of subpopulations is a random process. Therefore, for a sufficient number of, say, 20 patches, it is not very likely that all 20 subpopulations living on these patches will go extinct at the same time. Local extinctions occur independently of each other in an uncorrelated fashion, and therefore recolonization has the potential to counteract local extinctions.

However, this no longer holds if fluctuations in the biotic or abiotic environment, which increase the local extinction rate, affect larger parts of the landscape at the same time in the same way, i.e. are spatially correlated. Obviously, such correlations will increase the overall extinction risk of the metapopulation because now many patches may become unoccupied at the same time, leading to a lack of the emigrants required for recolonization. How strong the correlation of extinction events is depends on the type of main environmental fluctuations (e.g. bad weather, epidemics), the heterogeneity of the landscape, and the distance between pairs of patches. A mean correlation distance has to be specified for the META-X model.

4.5 Simulation

Once the user has specified a certain species and a certain landscape by defining all the model parameters, META-X uses the model to calculate the required quantities for risk assessment (mean time to extinction, extinction risk etc.). For this purpose, simulations are performed.

A simulation is a simulated population dynamics produced by the program which implements the META-X model on the computer. Starting with the initial occupancy pattern, the program calculates stochastically year by year which sub-population goes extinct and which is recolonized.

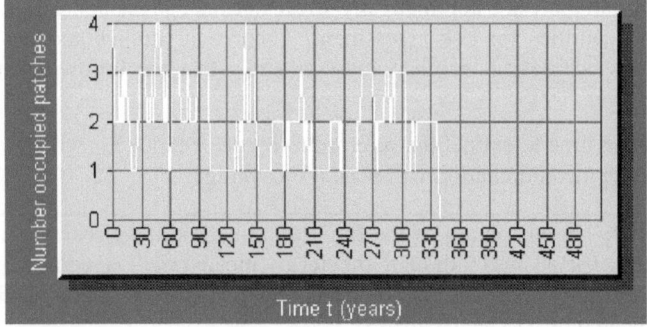

Typical result of a single simulation: number of occupied patches over time. At the end, the metapopulation goes extinct.

Control Parameters

A simulation ends if either the metapopulation goes extinct or if a certain time horizon, e.g. 300 years, is reached. To calculate these statistical quantities, META-X has (because of the stochastic nature of the modelled processes) to run simulations for a single set of model parameters many times (e.g. 1,000). The extinction times of the simulations then show a characteristic, theoretically predictable distribution which can be used to determine the intrinsic mean time to extinction and other quantities (see Chap. 13).

The time horizon and the number of simulations are referred to as control parameters in META-X.

META-X proposes standard values for these parameters which have proven to be useful in most cases, although users may change these parameters.

Interactive Simulations

Simulations may be run interactively, i.e. the user observes the population dynamics year-by-year and follows the local extinctions and recolonizations in detail. This mode of simulation serves to visualize and understand in detail the dynamics of the metapopulation and is not used when full risk assessments are required.

Automatic Simulations

Automatic simulations are much faster than interactive ones. As many landscapes, species and parameter sets as required are processed and the results presented.

4.6 Evaluation

The results of a simulation are automatically evaluated and presented. META-X calculates several quantities for risk assessment. Theoretical studies and experience with numerous population viability analyses show that the following three quantities are decisive:

1. The intrinsic mean time to extinction, referred to by the symbol T_m, which indicates how long the metapopulation will persist on average.

> **Important note**: The histogram of the time to extinction of individual simulations is asymmetric. Thus, although the term 'mean time' is correct in a mathematical sense, in most simulations the metapopulation will be extinct earlier than indicated by T_m (see Chap. 13).

2. The probability (or the risk) of going extinct within a certain time interval, referred to as $P_0(t)$ (pronounced: P-zero-of-t), where P indicates a probability, t the length of the time interval (e.g. 50 years) and the index 0 indicates extinction, i.e. a population size (number of occupied patches) of zero.
3. The probability of establishment or recovery, respectively, R_{ini}, of the initially specified metapopulation (Chap. 13).

4.7 Organizing Projects

META-X operates with three hierarchically organized units which – starting at the lowest hierarchical level – are explained in the following.

Scenarios

By specifying a certain species in a certain landscape, i.e. by specifying the complete set of model parameters, the user defines a scenario. A scenario is a specific ecological situation which can be simulated by META-X in order to assess the risk of metapopulation extinction and related quantities.

However, because of the inherent uncertainties in parameter estimation, individual isolated assessments of extinction are not very meaningful because they are more or less uncertain themselves. In order to go beyond isolated risk assessments which would suggest that these assessments are absolute, META-X supports the concept of comparative (computer-based) experiments where the focus is on relative risk assessments, i.e. on comparing assessments of different scenarios.

Experiments

An experiment is a set of one or more scenarios. Once an experiment consists of more than one scenario, it is automatically a comparative experiment, because experiments are units which can be completely simulated and evaluated by META-X.

Consider, for example, a case where the options are to remove a central, presumably important patch or to cut off the connections between three other patches. To analyse how the risk of extinction changes in these three scenarios (including the unaltered situation), these three scenarios should be defined as parts of one single experiment. Then you let META-X perform the (computer-based) experiment, i.e. run the simulation and evaluate the experiment. As a result, comparative diagrams (and tables) are produced which compare the mean time to extinction (or other quantities) for the three scenarios.

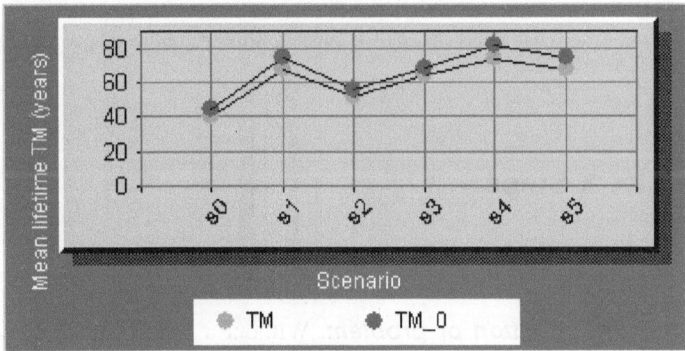

Typical result of a comparative experiment of META-X: Comparison of the mean time to extinction (TM) of different scenarios.

Instead of different configurations of the landscape, different scenarios may also represent different values of a single model parameter. In this way, comparative experiments can be used to determine the sensitivity of the mean time to extinction to small (and large) changes in model parameters.

Projects

Experiments which belong to each other because, for example, they address the same species or the same landscape, are grouped together in projects. Projects are the units which are saved to and loaded from your hard disk by META-X.

A project is like a file. You are free to choose the criteria for defining and organizing these files (although for reasons of management of the computer memory it would not be a good idea to have too many experiments in one project). META-X only deals with the structure within these files (projects). The structure of the project being worked on is visualized all the time in the project window containing the Project Tree.

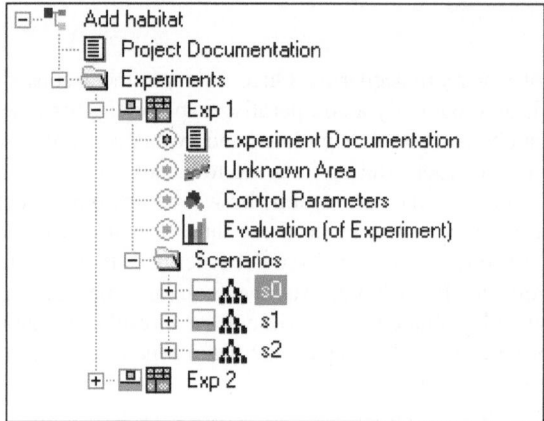

Typical project tree, divided into project name (Add habitat), two experiments (Exp 1, Exp 2), and three scenarios (s0, s1, s2).

4.8 The META-X Steps

While working with META-X you will always have to perform the following five basic steps:

1. *Formulating the question or problem*. Without a clearly stated question you will not be able to work with META-X (or any other model). However, META-X may help you refine your question afterwards because the answers to an earlier question may lead to a better understanding of what is really essential in order to ask (and to answer) the question.
2. *Defining scenarios*. By choosing the model parameters, you focus META-X on a certain species in a certain landscape. For real species and landscapes, this step is the most difficult because the database needed to specify the parameters is generally incomplete.
3. *Defining experiments*. You define comparative (computer-based) experiments which are designed to provide answers to your questions. You do this by defining several different scenarios and grouping them in an experiment. Likewise, uncertainties in parameter estimation are taken into account by varying individual model parameters and grouping the resulting scenarios in experiments which could then be called 'variation experiments'.
4. *Performing and evaluating experiments*. Experiments are automatically simulated and evaluated by META-X.
5. *Critically interpreting the results*. Although META-X will not interpret the results for you, your interpretation is supported by comparative diagrams.

If you want to use META-X for problems of landscape planning or biological conservation, a further step may be required:

6. ***Making decisions***. Again, although META-X cannot take decisions for you, it is designed to aid your decisions, which may affect the persistence of a meta-population.

Chapter 16 presents examples of different problems and questions from different areas and demonstrates how to perform the various steps of META-X. The entire META-X book will help familiarize you with these steps.

5 Guided Tour

Goals

In this chapter, the most important elements of working with META-X are demonstrated in step-by-step instructions. Make sure you take the Guided Tour on your computer in order to understand what META-X is and how it works. You will learn:

- How to define a project.
- How to define an experiment.
- How to define your first scenario.
- How to perform your first interactive simulation.
- How to define a second scenario which is based upon your first scenario.
- How an experiment is automatically simulated.
- How to interpret and evaluate your results, and how to export them.

5.1 Your First META-X Experiment

Install META-X as described in Chapter 3 or, if META-X is already installed, start META-X.

Defining a New Project

After you have started META-X, the META-X window appears:

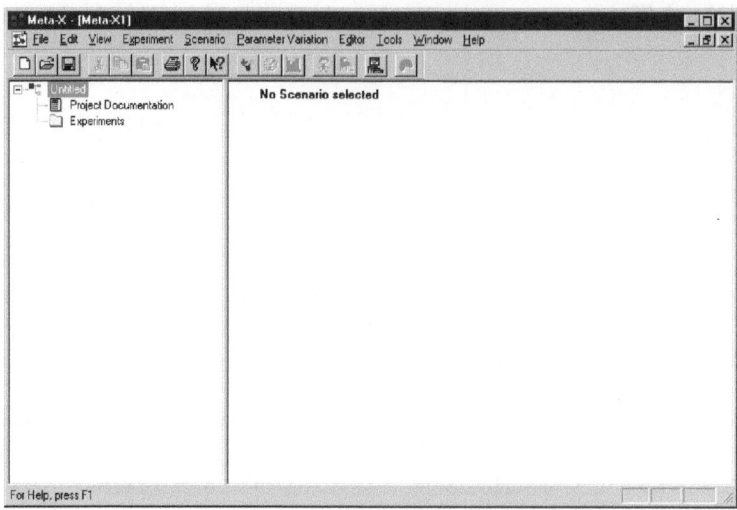

The META-X window consists of the following components, which are described and explained below:

- **Menu** (top)
- **Tool bar**
- **Project window** with Project Tree (left) and the window of the Landscape Editor (right)
- **Status bar** (bottom)

When started, META-X opens an 'empty' project which serves as a template for defining a new project. You can make this empty project your first project by saving it under a new name:

Creating a New Project by Saving

1. In the **File** menu, choose **Save As**.
2. In the **Save As** dialog box, enter in the box 'File name' the file name 'First Project'.
3. Click **Save**.

Note: When using the **Save As** dialog box, you can, of course, use all the procedures provided by Windows for saving files, e.g. changing directories or creating new ones, etc.

The META-X window should now display the project's name 'First Project' in the title bar. The newly created project file is named 'First Project.mtx'.

Enter Project Documentation

1. In the project window, double-click in the project tree on the text (or its icon) **Project Documentation**. The following window opens:

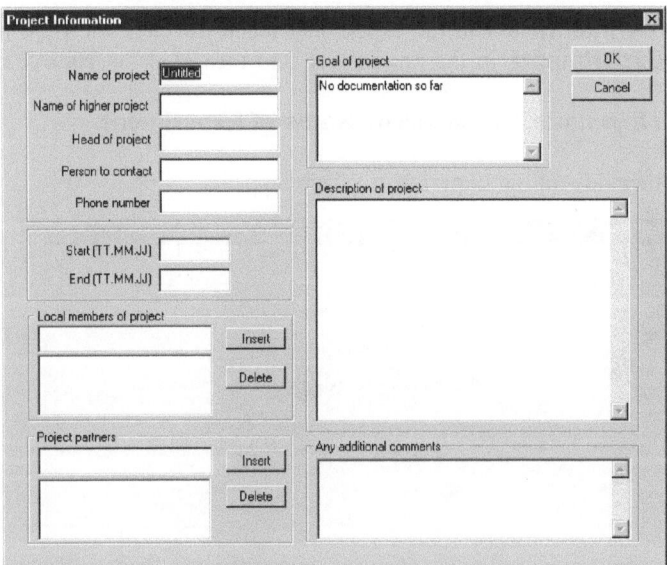

2. In the field 'Name of Project', enter 'First Project'.

Note: You can move forwards from box to box in this window by using the tab key (or for reverse movements the shift+tab key), or you can move to the individual boxes by clicking them.

3. In the field 'Start', enter the date on which you start the project (using the format 01.05.02 for May 1, 2002).
4. In the field 'Goal of (Sub-)Project', enter the goal or the problems and questions addressed in this project, e.g. "First Steps with META-X".
5. In 'Description of (Sub-)Project', enter the subject of the project, e.g. "First project for learning how to use META-X (from Chap. 5 in the META-X book)". Note that lines cannot be wrapped manually in the text field of the project documentation.

Note: In your later work with META-X, try to make good use of the project documentation because it will help you organize your work.

The project documentation you enter is automatically included in the report about the whole project which can be produced automatically by META-X.

6. In the box 'Local Members of Project', enter your name.
7. Click **Insert**. Your name will be included in the name list.

8. Click **OK**.

Creating and Documenting Your First Experiment

1. Choose **Experiment|New** in the experiment menu (or click the right mouse button on **Experiments** in the project tree and then click on **New Experiment**).
2. Enter "First Experiment" in the window **Name of Experiment**.
3. Click **Next**.
4. The control window of the Experiment Wizard appears:

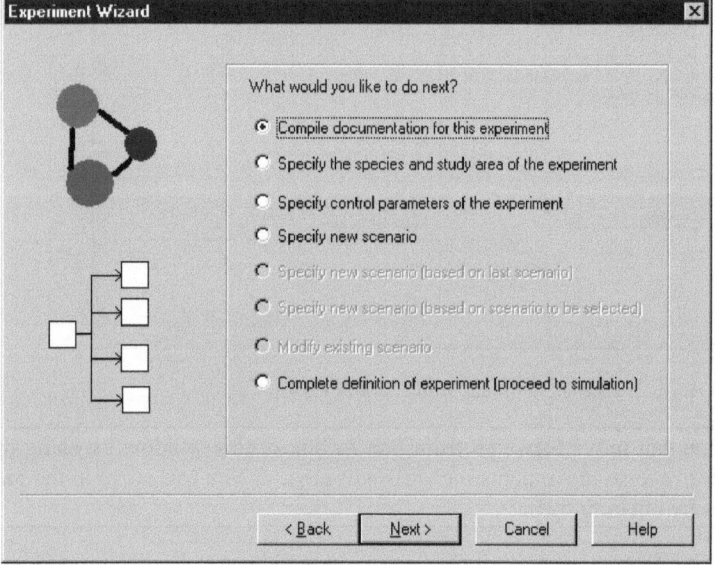

It shows the list of job steps required for the definition (and modification) of experiments. In this list, the next step that will be performed if you click **Next** is tagged. You may, however, use other steps by clicking on the corresponding item in the list.

5. Click **Next**.
6. In the window 'Documentation of Experiment', enter your name in the box 'Author', and in the box 'Please enter description' type: "First Experiment: A metapopulation of five patches (from Chap. 5 in the META-X book)".
7. Click **Next**.

Again, the control window of the Experiment Wizard appears. Now the step 'Specify Species and the Study Area of the Experiment' is highlighted and is executed next:

Naming the Species and Delineating the Landscape

1. Click **Next**.
2. In the window **Study Area**, enter "Hypothetical species" into the box 'Name of the species' and "Hypothetical landscape" into the box 'Name of the study area'.

Next you may choose the spatial unit used by META-X. 'Km' is default. We will leave this unchanged.

The next box is 'Position (Coordinates)'. Here, the coordinate system is specified by which you will later define the coordinates of the patches:

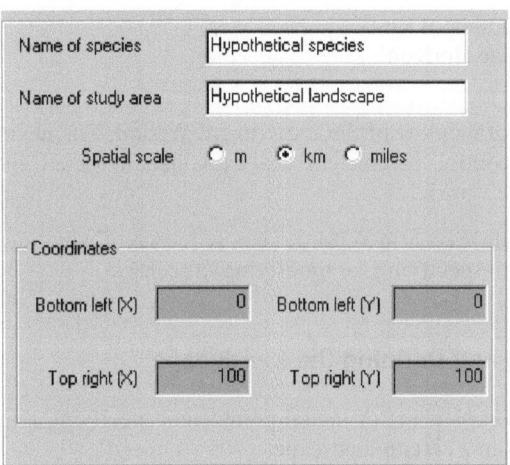

Note: In META-X, input fields which are currently empty or which are filled with invalid values have a red background; validly filled fields have a green background.

---

3. Enter 0 for the 'Lower' X and Y coordinates and 100 for the 'Upper' coordinates, i.e. overwrite the '?' in the input fields. You can use tab-keys or direct positioning by mouse clicks to navigate between the fields.

Now you have defined a quadratic clipping of the landscape with a size of 100 × 100 km². The lower, left-hand corner is (0,0) and the upper right-hand corner (100,100).

4. Click **Next**.

In the control window of the Experiment Wizard, the next step is now 'Specify control parameters of the experiment'. Here you have to specify those (control) parameters which control your simulations.

Defining Control Parameters

The following control parameters have to be specified:

Control parameter	Meaning
Number of Runs	How many simulations are to be run
Time Horizon	Maximum duration of a single simulation (in years)

We will use control parameters which have usually proved adequate:

1. Enter '1000' in the box 'Number of Runs'.
2. Enter '1000' in the box 'Time Horizon'.
3. Click **Next**.

We are now back in the control window of the Experiment Wizard. The next step is 'Specify new scenario' (control parameters are explained in detail in the Chaps. 7 and 13).

4. Click **Next**. This will start the Scenario Wizard, i.e. a sequence of windows and dialog boxes which assists in specifying (or modifying) a scenario.

Scenario: Documentation and Defining the Landscape

Remember? A scenario is a complete set of model parameters specifying a metapopulation of a certain species in a certain landscape.

1. In the window 'Documentation of Scenario' type 's0' in the Box **Name**.
2. Type a text such as: "In this basic scenario, a metapopulation living on five patches is defined (see Chap. 5 in the META-X book)" in the box 'Please enter comment'.
3. Click **Next**.

Note: You can also use the names which are automatically assigned to new scenarios by META-X. These names are composed of the experiment's name and a number.

4. Enter the number 5 in the window 'Number of Patches for a Scenario'.
5. Click **Next**.
6. In the window 'Position of Patches', enter the X and Y coordinates (position of patches) of the five patches exactly as described in the following screen shot:

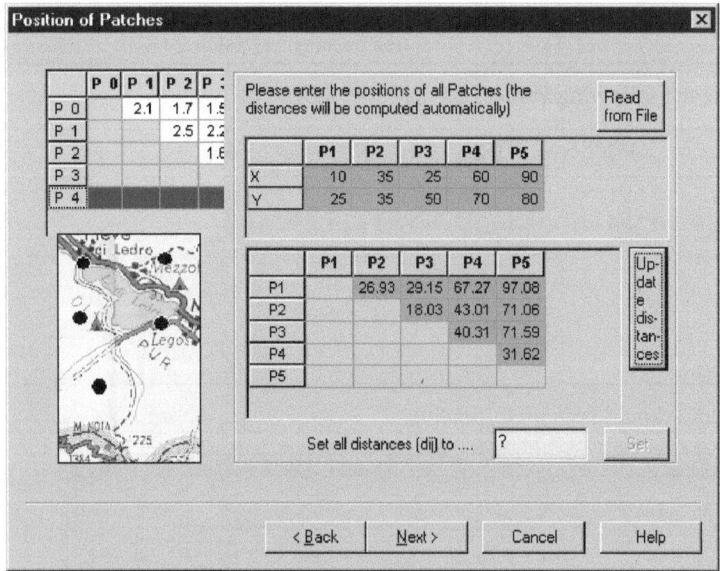

7. In the table of coordinates, you may move from box to box by pressing the tab key or by mouse-clicks on the boxes.

Note: If you try to enter a coordinate which is beyond the borders of the coordinate system defined earlier, the background of the coordinate box turns red.

8. Click **Update Distances**.

Now the distance between each pair of patches is calculated and displayed in the distance matrix. The boxes 'set all distances (dij) to...' and 'Read from File' are explained in Chap. 7.

9. Click **Next**.

Patch Characteristics

1. In the window 'Local Characteristics of the Patches', you are asked to enter the following for each of the five patches:

Parameter	Meaning
Ext. Rate	Probability per year that a subpopulation inhabiting this patch goes extinct.
Emigrants per Year	Average number of emigrants per year which are produced by this patch.
Immigrants Needed	Number of immigrants which are needed in a certain year to establish – with a

	probability of 50% – a new subpopulation on an empty patch.

Please use the following values:

	P1	**P2**	**P3**	**P4**	**P5**
Prob. Ext.	0.1	0.05	0.05	0.001	0.1
Emigrants per Year	1.7	2.1	1.3	17	5
Needed Immigrant	8	5	4	4	10
Occupancy	✓	✓	✓	✓	✓

The window should now look like this:

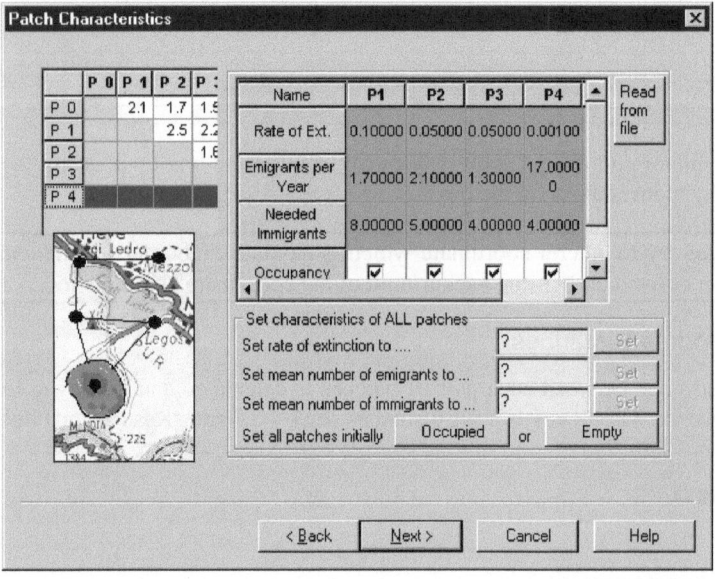

Note: Patch P4 is relatively safe (mean time to extinction: $1/0.001=1,000$ years) and produces many emigrants. Patches P1 and P5 persist for only 10 years on average and need many immigrants to become recolonized. You should always try to 'read' model parameters in this way, i.e. to translate them into terms which have a direct meaning concerning metapopulation dynamics and extinction risk.

Please note the default of META-X that all patches are occupied at the beginning of a simulation. Later, in your own applications, you can change this by clicking the checkboxes which indicate occupancy.

2. Click **Next**.

Connection Between Patches

Here you have to specify the pairs of patches between which recolonization may occur:

1. In the window 'Connections of Patches' click **Connect all**.

Now all the checkboxes which indicate whether a connection exists between two patches are marked.

2. Deactivate the connections between patches P4 and P3 and P5 and P1 by clicking the corresponding checkboxes.

The connectivity matrix should now look as follows:

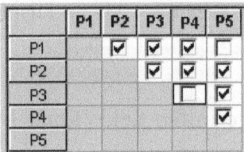

Here, a 'matrix' is simply a convenient way to arrange all combinations of two different patches.

3. Click **Next**.

Correlation Length

In the window 'Correlation Length' you are now asked to specify the mean correlation length. There are two alternative ways to do this: either by defining the mean correlation length directly or by specifying the probability that extinction events on two different patches are correlated, i.e. occur simultaneously. META-X then calculates the mean correlation length from this sample correlation between the two patches.

For the Guided Tour we will choose the direct method (for an explanation of the meaning of the parameter 'mean correlation length', see Chaps. 14 and 15):

1. Click **Next**.
2. In the window 'Correlation Length (Direct)', type in the field 'Enter Mean Correlation Length (d0)' the value 25.
3. Click **Reset Matrix**. A message appears warning that a 'reset' may overwrite earlier manual changes in the correlation matrix. Ignore this warning by clicking **Yes**.

The correlation matrix should now appear as follows:

	P1	P2	P3	P4	P5
P1		0.34055	0.31161	0.06783	0.02058
P2	0.34055		0.48617	0.17899	0.05829
P3	0.31161	0.48617		0.19941	0.05706
P4	0.06783	0.17899	0.19941		0.28230
P5	0.02058	0.05829	0.05706	0.28230	

4. Click **Next**.

Dispersal Range

As with the mean correlation length, you can choose between two alternatives to specify mean dispersal range, i.e. direct or indirect (window 'Mean Dispersal Range'). Please choose the direct method:

1. Click **Next**.
2. In the window 'Mean Dispersal Range (direct)' type in the field 'Enter Mean Dispersal Range (d1)' the value 45.
3. Click the button **Reset Matrix**.
4. Ignore the warning message by clicking **Yes**.

Now the reachability matrix has been calculated. The entries in this matrix indicate the principal probability that an emigrant from patch i will reach patch j if there is a connection between these two patches. Thus, reachability is calculated from the patch distance and the mean dispersal range, but ignores the connection matrix.

The reachability matrix should now read as follows:

	P1	P2	P3	P4	P5
P1		0.54967	0.52321	0.22427	0.11563
P2	0.54967		0.66987	0.38451	0.20616
P3	0.52321	0.66987		0.40829	0.20374
P4	0.22427	0.38451	0.40829		0.49526
P5	0.11563	0.20616	0.20374	0.49526	

5. Click **Next**.

The window 'Rates of Colonization' appears and shows the colonization matrix, which is calculated from the reachability matrix, the connection matrix as well as the patch characteristics 'Emigrants per Year' and 'Number of immigrants needed'. The entries in the matrix indicate the rate of colonization between each pair of patches. Note that some of these entries are zero because the corresponding connections do not exist:

	P1	P2	P3	P4	P5
P1		0.03115	0.03706	0.01589	0.00000
P2	0.01804		0.04396	0.02523	0.00541
P3	0.01417	0.02903		0.00000	0.00441
P4	0.07943	0.21789	0.00000		0.14032
P5	0.00000	0.03436	0.04245	0.10318	

6. Click **Next**.

Main Model Parameters

The window 'Main Model Parameters' now presents an overview of the main model parameters, i.e. those parameters which are directly used by the model to simulate metapopulation dynamics. Note that you entered some of these parameters directly, whereas others are calculated from parameters you entered in submodels:

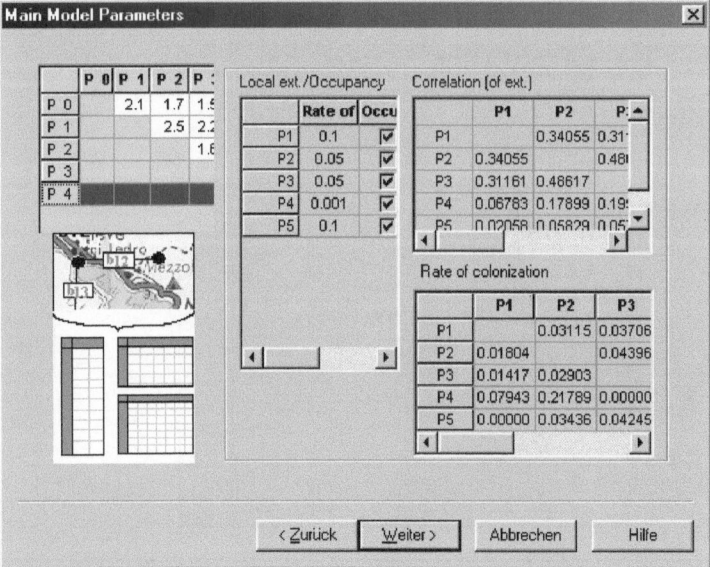

The main model parameters are:

- **Local extinction rates** and **occupancies**. These parameters were entered directly.
- **Degrees of correlation**. These were calculated from the patch distance and mean correlation length
- **Colonization rates**. These were calculated from patch distances, the connectivity matrix, the 'Number of emigrants', the mean dispersal range (via the reachability matrix), and the 'Number of immigrants needed'.

Note: You may also directly specify the reachability, correlation or colonization matrix or modify entries in these matrices (see Chaps. 7 and 13 for a detailed description of the hierarchical structure of the META-X model and Chap. 15 for the issue of parameterization).

1. Click **Next.**

The window 'Control Parameters for the Whole Experiment' now allows us to tailor some or all control parameters to the current scenario which may differ from the control parameters specified for the whole experiment. We will not do this here (and in most cases it will anyway not be necessary).

2. Click **Next.**

In the next window 'End of Scenario Specification' you may decide to click **Back** to check your definition of the scenario and possibly to modify it, or you may click **Next** to proceed to the Experiment Wizard where you can define further scenarios, modify existing scenarios, or complete the definition of the experiment and proceed to simulation.

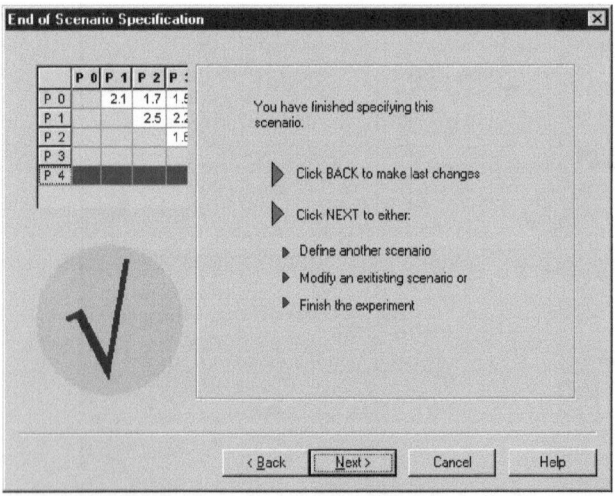

Note: The **Back** button is intended only for navigation in the wizard. It has no 'undo' functionality.

3. Click **Next.**

The program returns to the Experiment Wizard. The next step in the wizard 'Specify new scenario (based upon last scenario)' would be to define a new scenario based on the scenario you just defined. Here we will instead finish the experiment wizard and proceed to interactive simulation:

4. Click the item 'Complete definition of experiment (proceed to simulation)'. You can see by the black dot next to this item that this item has now been activated as the next step.
5. Click **Next**.

In the window 'End of Experiment' you can choose to proceed to the Simulation Wizard immediately or, if you prefer, to return to the main program:

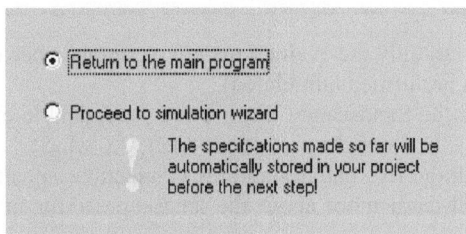

6. Since we want to save our work so far, we return to the main program. Click **Return to the main program**.
7. Click **Next** and in the following window 'Overwiew of Experiment', read the message and then click **Finish**.

You can now see in the Project Tree (left part of the META-X window) the new experiment 'First Experiment'. If you click on the plus icon in the tree, you can see the components of the experiments as well. You can also do this with the plus icon of the scenario:

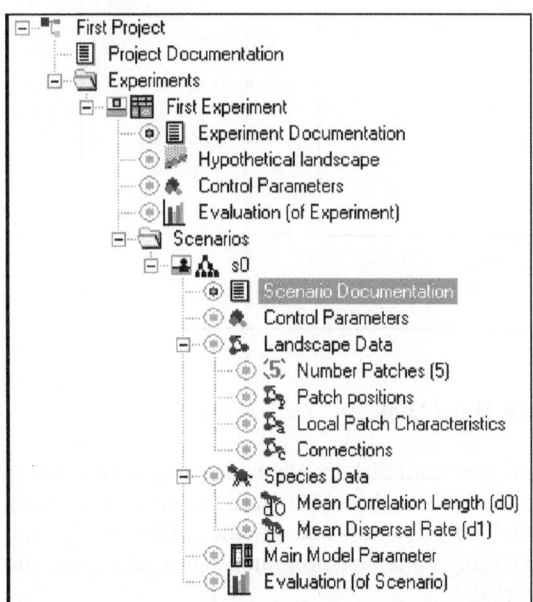

Important note: The icons, i.e. the squares and circles, linked to the experiments and scenarios, indicate whether the corresponding units are ready to be simulated, i.e. whether the main model parameters and control parameters have been completely specified. Green indicates 'ready for simulation', red (as a warning color) shows 'not ready'. Likewise, the colour of the circles linked to the components of experiments or scenarios indicates whether the components have been completely specified (green) or not (red).

In the project you have defined so far, only the evaluations are incomplete because the experiments have still not been performed (simulated).

To obtain an initial idea of what the Landscape Editor is and does, double-click on the name of the scenario ('s0'). In the right part of the META-X window, the landscape you defined appears with the five patches and the connections specified in the connection matrix. You will learn more about the landscape editor in the Chapter 9.

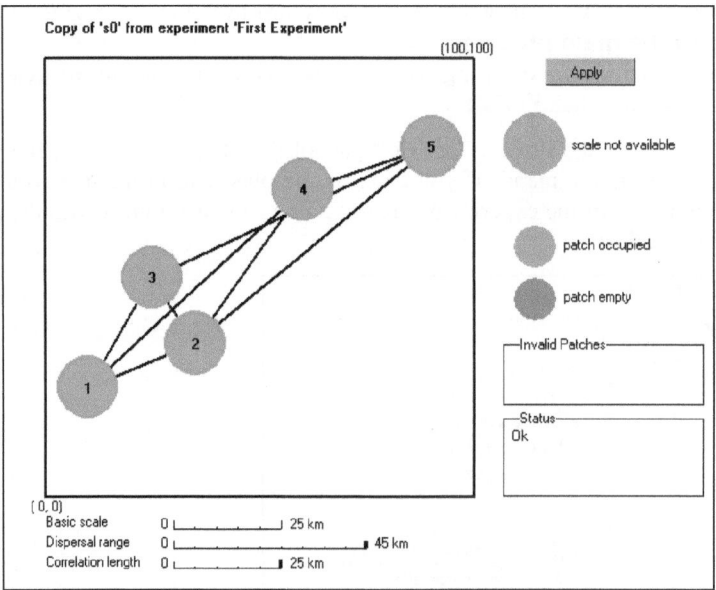

Saving the Project to Your Hard Disk

Please now save the project to your hard disk by choosing **File|Save** in the menu or by clicking the diskette icon in the toolbar. Your work so far is now saved and you could interrupt the Guided Tour for now and resume it later by opening the project 'First Project'. As with any computer program, it is a *very* good idea to save your work at regular intervals!

5.2 Interactive Simulation

Due to the stochastic nature of extinction and recolonization, the basic quantities calculated by META-X to assess extinction risk – mean time to extinction, risk of extinction within a certain time interval, and patch incidence (i.e. probability of a patch being occupied by a subpopulation) – are statistical quantities which are gained from many (typically 1,000) simulations. However, these statistics do not make META-X a black box where the user simply has to 'believe' the output. Instead, META-X allows simulations to be run interactively. You can thus follow one time step at a time to see how patches go extinct and are recolonized depending on the landscape structure and the main model parameters. This enables you to acquire an intuitive feel for and an understanding of the dynamics of the metapopulation because you will 'see' the significance of each patch and each connection between patches for overall persistence.

Note: Interactive simulations are unsuitable for complete evaluations of scenarios or experiments because in interactive mode simulations run much too slowly. For complete evaluations, simulations must be run in automatic mode (see below).

Choosing a Scenario

1. If you interrupted the Guided Tour, load your project 'First Project' with **File|Open**.
2. Choose **Scenario|Interactive Simulation**.
3. In the window 'Select Experiment and Scenario', select in the box 'Select Experiment' the experiment 'First Experiment' by clicking on the experiment's name.

The box 'Select Scenario' displays the list of scenarios which have hitherto been defined for the current experiment. We have defined only one scenario so far, i.e. 's0'.

4. Select this scenario by clicking on the name of the scenario and then clicking **OK** or by double-clicking the name:

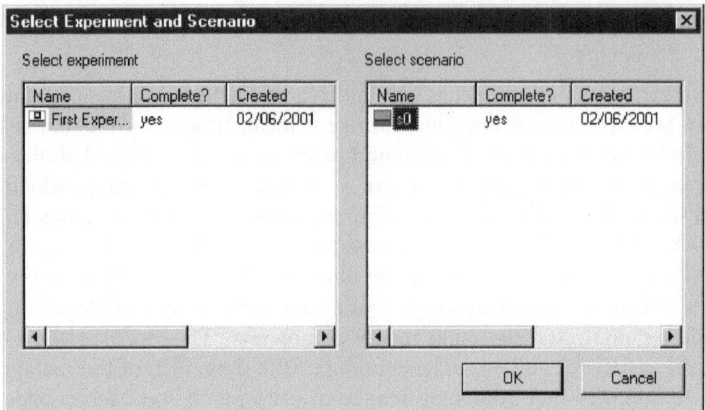

The control window of the interactive simulator appears ('Interactive Simulation of Scenario 'First Scenario''). All features of the interactive simulator are accessible from this control window:

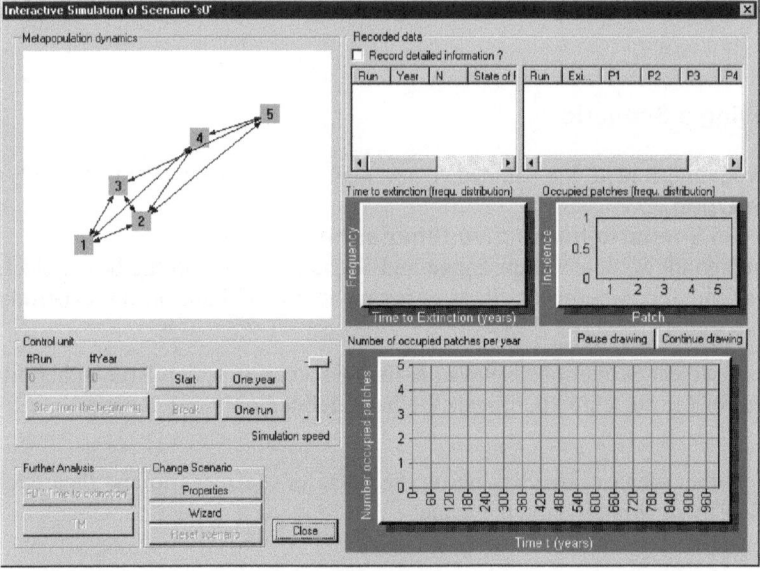

The control window has the following panels:

- **Metapopulation dynamics**. Here the landscape defined in your experiment and the scenario are visualized. Patch icons in blue indicate occupied patches, icons in grey unoccupied patches (i.e. the corresponding subpopulation is extinct).
- **Recorded data**. If the checkbox 'Record detailed information' is active, the following information is recorded in the left window of this panel: number of simulation run, current year, number of patches occupied in this year and the

occupancy of the entire metapopulation ('1' indicates an occupied patch, '0' an empty one). The right window shows for each simulation the year of extinction (or the year when simulation stopped, i.e. the time horizon), and the probability of patches being occupied (incidence).

- **Time to extinction (freq. distribution)**. A diagram showing the frequency distribution (histogram) of the times to extinction.
- **Occupied patches (freq. distribution)**. A diagram showing the incidence of the patches.
- **Number of occupied patches per year**. A diagram showing the time series of the number of occupied patches.
- **Control unit**. Here you can control the interactive simulation (see below).
- **Further analysis** (see below).
- **Change Scenario**. Here you can (assuming you have interrupted the simulation) change the control parameters (**Properties**) or the model parameters (**Wizard**). Once you are done with interactive simulation, you will be asked whether you want to save these changes and thereby modify your original scenario.

Interactive Simulation

You can play around with and explore the interactive simulator as much and for as long as you want. In the following we will give just a few examples:

1. Activate the checkbox 'Record detailed information'?.
2. Click **One year**, say, ten times. While doing this, observe the landscape in the panel 'Metapopulation dynamics'.
3. In the window 'Recorded data', move the scroll bar of the left table to the right until you see the full list of 'State of the Patches':

Run	Year	N	State of
0	0	5	1,1,1,1,1
0	1	4	1,0,1,1,1
0	2	5	1,1,1,1,1
0	3	5	1,1,1,1,1
0	4	5	1,1,1,1,1

4. Continue clicking **One year** and observe how the pattern of occupied ('1') and empty ('0') patches changes.

Check Parameters

5. To check whether this result is consistent with the parameters you specified, click **Properties**.

A dialog box appears with numerous tabs displaying all the parameters of the experiment and scenario:

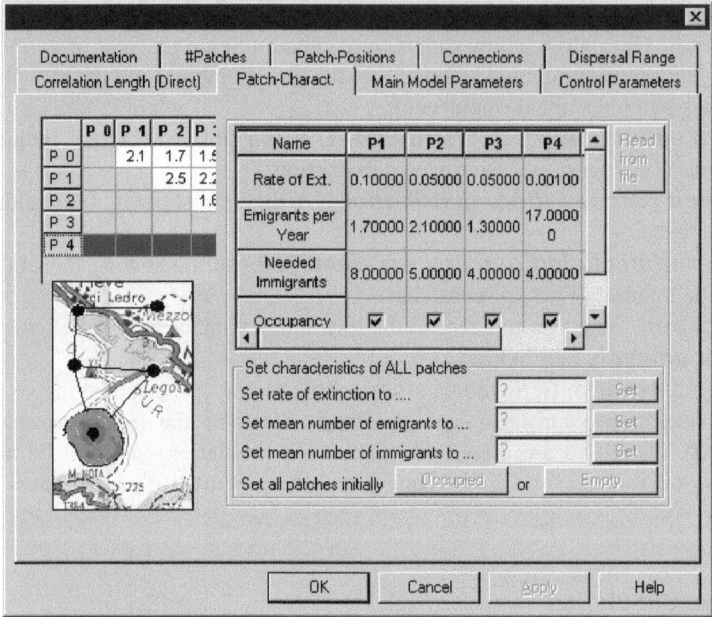

6. For example, click the tab 'Patch-Charact.' and check if the simulation results are consistent with the parameters of the patches.
7. Click **Cancel** to return to the simulator.
8. Click **One run**. The current simulation is now run automatically until the metapopulation is extinct or until the time horizon specified in the control parameters (here: 1000 years) is reached.

Pattern of Occupancy

9. Let this simulation run until the end. Then you can study the temporal dynamics of the occupancy (column: 'State of Patches') by using the vertical scroll bar.

The diagram 'Number of occupied patches per year' shows the time series of the number of occupied patches:

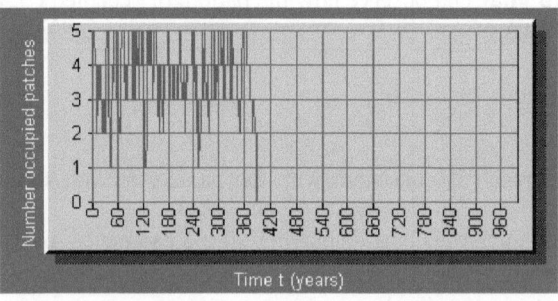

Note: The time series will very probably look different on your screen because 'your' simulation is likely to have used different random numbers.

Now run several simulations by clicking **One run** several times. Keep in mind that in each simulation run different random numbers are used so that no single simulation looks exactly the same as any other simulation. You can also see this in the frequency distribution 'Time to extinction', which changes after every simulation. The reason is that the times to extinction vary from simulation to simulation and thereby change the frequency distribution.

To interrupt simulations, click **Break**, to continue click **Continue**. Clicking **Start from the beginning** deletes all the results obtained so far in the interactive simulation and starts again from the beginning (run 1, year 1).

For a statistically sound evaluation, the interactive mode of simulation is much too slow. Nevertheless, you can try to analyze the significance of certain patches (or connections, or parameters etc.) for metapopulation persistence by observing the incidence, i.e. the probability that a patch was occupied during the last run that was finished:

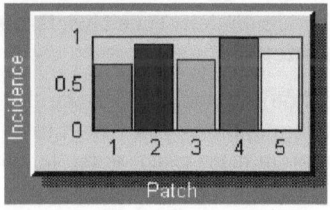

Modifying the Scenario

1. You can explore the significance of certain patches by interrupting the simulation with **Break** and then clicking **Wizard**.
2. In the Scenario Wizard, click **Next** until you can change the number of patches. Then in the input field 'Please enter number of patches', replace 5 by 4.
3. Click **Next** and then in the window 'Select patches to be removed' click on P4 and then on the button with the arrow pointing to the right:

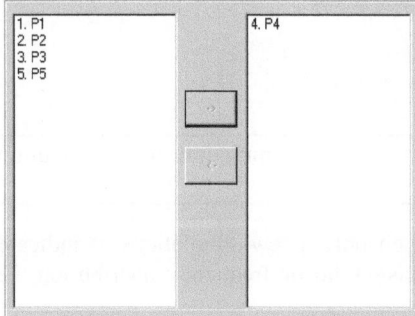

4. Click **OK**, **Next** or **Yes** until you reach the window with the button **Finish**. Click **Finish** to return to the simulator.

Note the changes in the landscape: the original patch P4 and all connections to P4 have disappeared.

Note: In META-X, patches are always numbered consecutively. Therefore, in the modified scenario there is still a patch P4, even though this was patch P5 in the original scenario.

Mean Time to Extinction

In the following, we are going to obtain an initial, rough estimate of the mean time to extinction of the metapopulation in the modified scenario. We will use procedures grouped in the box 'Further analysis'.

1. Click **Start** and let – assuming you have the time – the simulator run for, say, 25 simulations. **Pause drawing** will speed up simulation. Then click **Break**.
2. In **Further analysis** click **FD "Time to extinction"**.

The window 'Time to Extinction (Freq. Distribution)' appears displaying the frequency distribution of extinction times from the simulations you have run so far (e.g. 25). The default 'number of years per histogram class (bar)' is the length of the time horizon divided by 50 (here: 1000/50=20):

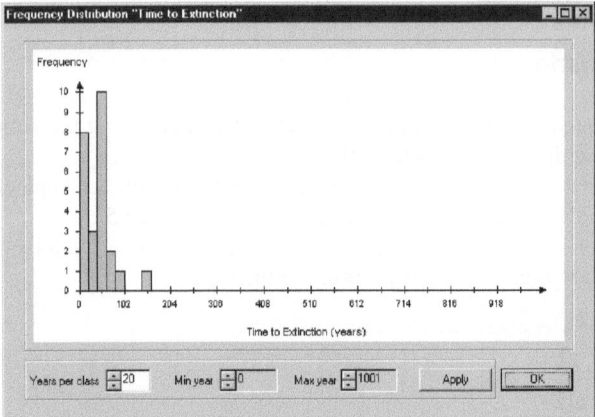

Note: Because of the random processes in the simulation, this distribution will look different on your screen.

Even though this histogram is based on only a few simulations, it indicates the typical exponential decrease of the histogram or frequency distribution, i.e. the probability of observing extinctions in a certain time interval decreases exponentially in the course of time.

3. Click **OK**.
4. In 'Further Analysis' click **TM**.

In the window 'Compute TM from the frequency Distribution 'Time to Extinction'' you can again see the histogram of extinction time in the small inlet in the upper right-hand corner.

5. Click one after the other on **P0(t)**, **Relevant P0(t)**, **-ln(1-P0(t))** and **TM**:

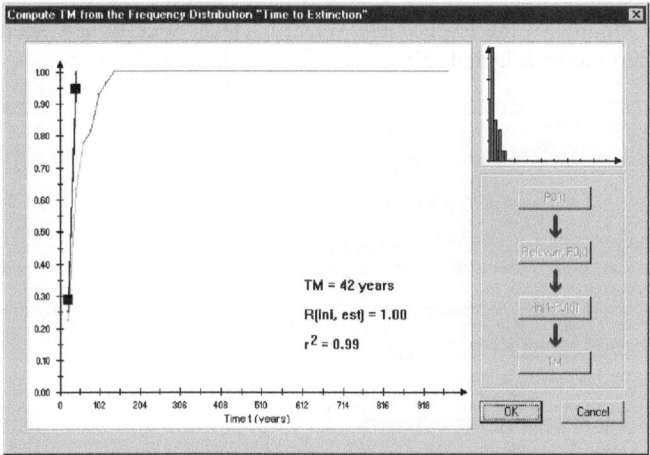

Note: By clicking these four buttons, the four steps for calculating the intrinsic mean time to extinction, T_m, are executed. For an explanation, see Chapter 13.

After clicking **TM** you can see in the diagram the calculated value of T_m (e.g. 42 years) and the probability of establishment, R_{ini} (e.g. 1). Note that with about 25 simulations these assessments are usually rather imprecise. This is why the control parameter 'number of runs' is usually set to 1000.

6. Click **OK** to return to the interactive simulator.

You can – and should – try to perform further analyses with the interactive simulator, e.g.: how does T_m change if you run more and more simulations? How large is T_m for the original scenario (you can return to this by clicking **Reset Scenario**)?, etc.

We will instead now leave the interactive simulator, define a second scenario and then start an automatic simulation.

7. Click **Close,** then **Yes** and then **No** (i.e. you do not want to save the modifications to the scenario 's0' that you made during interactive simulation).

Adding a Second Scenario

The analyses started during interactive simulation are now going to be specified and evaluated more systematically. First, we add a second scenario to the experiment 'First Experiment'.

1. Choose **Experiment|Modify**.
2. In the window 'Select Experiment' select the experiment 'First Experiment' (by clicking on the experiment's name). Click **OK**.
3. In the window 'Name of Experiment' click Next because we do not want to change the name.
4. In the control window of the Experiment Wizard select 'Specify new scenario (based on scenario to be selected...)':

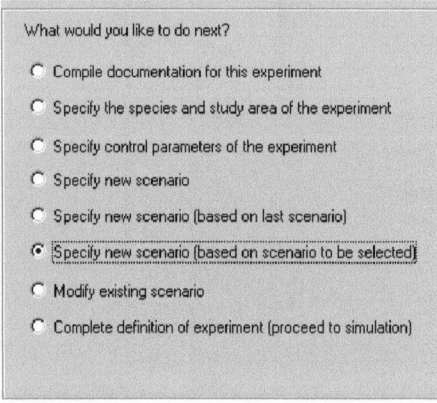

5. Click **Next**.
6. In 'Choose scenario to be modified' there is only one scenario ('s0') to select. Therefore, click **Next**.
7. In 'Documentation of Scenario' enter 's1' in the field 'Name', and add a text like "Like first scenario, but patch #4 deleted." in the field 'Please enter comment'.
8. Click **Next**.
9. Enter '4' for the number of patches, then click **Next**.
10. Select patch P4 and click on the button with the arrow pointing to the right. Click **OK**.
11. Click **Next** or **Yes** or **Finish** until you have returned to the control window of the Experiment Wizard.
12. Select 'Complete definition of experiment (proceed to simulation)' and **Next**.
13. Select in the next window 'Return to the main application'. Click **Next** and then **Finish**.

If now in the project tree you click on the plus icon near 'Experiments' and then on the plus icon near 'Scenarios', you will see that the experiment 'First Experiment' now contains two scenarios:

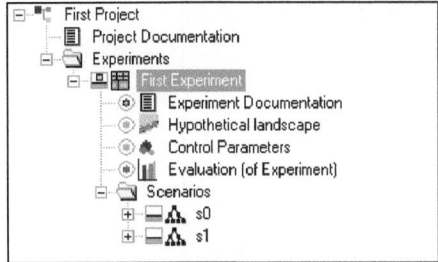

If you click the '+' belonging to 's1', then 'Landscape Data' and finally 'Number of Patches', you will see the number of patches modified to 4. A double-click on the identifier 's1' in the Project Tree makes the Landscape Editor display the new scenario, with only four patches.

14. To save your work so far, choose **File|Save** in the menu.

5.3 Automatic Simulation

1. Choose **Experiment|Simulate...** .
2. In 'Select Experiment and Scenarios' there is nothing to select because there is only one experiment.
3. Select both scenarios: First, click on 's0', then press the Shift or Ctrl key and click on 's1'.
4. Click **Next**.

The next window shows the control parameters for the scenarios selected.

5. Click **Next**.

The following window appears:

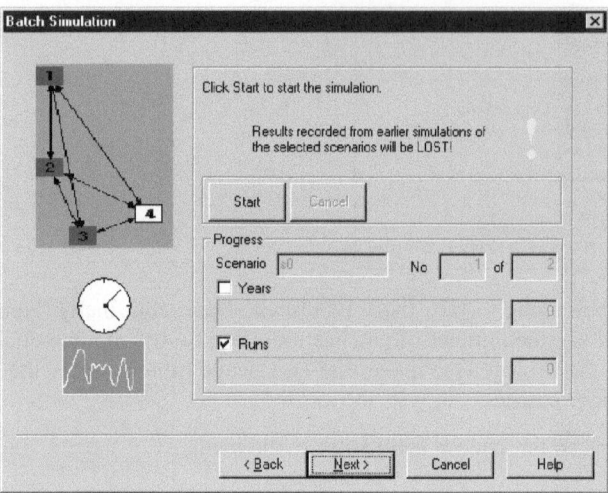

6. Click **Start**.

In the panel 'Progress', the field 'Scenario' shows which scenario is currently simulated, while the fields 'No' and 'of' indicate the number of the current scenario and the total number of scenarios to be simulated.

You will see that in this example automatic simulation is at least 100 times faster than interactive simulation.

7. In the message box which informs you about the end of the simulation, click **OK** and then **Next**.

Automatic simulation is now finished. Easy, wasn't it? The raw data obtained from the simulation now have to be evaluated statistically. This next step of META-X is called evaluation.

Note: The raw data are not permanently stored. If you click **Cancel** now, the primary data will be lost.

5.4 Evaluation

In the window 'What Do You Want To Evaluate' you can select the scenarios for which you want to evaluate the raw data. Default is to evaluate all scenarios, i.e. 's0' and 's1'.

You can also choose which output quantities you want to be calculated. Again, we leave the default unchanged.

1. Click **Next** to start evaluation.

The window 'Evaluation' appears showing the results of the automatic simulation:

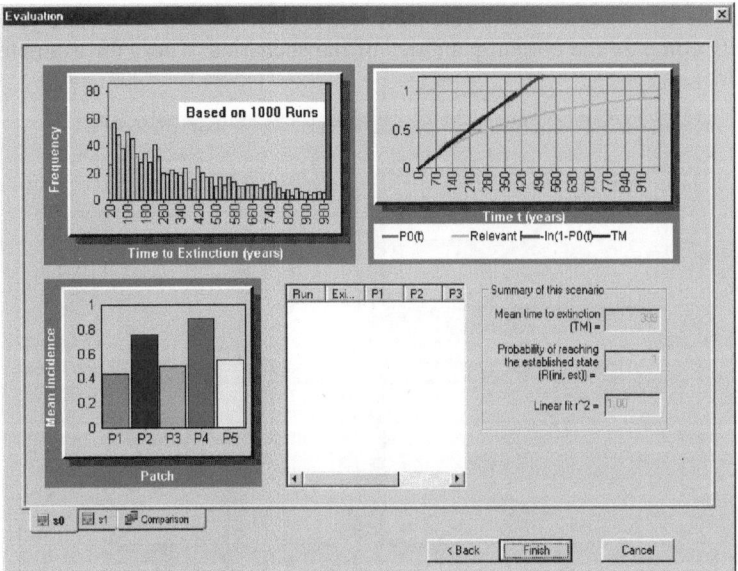

There are three tabs ('s0', 's1', 'Comparison').

Note: You can copy the diagrams displayed on the evaluation window to the clipboard or print them directly via a context menu: click the right mouse button while the cursor is placed over the diagram. If you place the cursor over a certain point or bar in the diagram and wait for a while, the number of the value on the x-axis and its value are displayed.

The tabs of the scenarios show those diagrams which you already know from interactive simulations:

- The frequency distribution (histogram) of the times to extinction.
- A diagram showing $P_0(t)$ and the $-\ln(1-P_0)$-plot.
- The mean incidence of the patches, i.e. the probability that a certain patch is occupied in a certain year.
- A table containing data of each individual simulation. This table is empty because we decided earlier not to record detailed data.
- A 'Summary of this Scenario', i.e. an overview of the most important output variables, T_m and R_{ini} (and r^2, the regression coefficient of the linear regression used in the $-\ln(1-P_0)$-plot).

2. Compare the two scenarios by clicking on their tabs.

We recall that patch P4 of the first scenario was deleted in the second scenario.

The result of this first experiment is that for a metapopulation whose mean time to extinction is anyway rather small (about 400 years), removing patch P4 would further dramatically reduce the mean time to extinction (to about 30 years).

3. Select the tab 'Comparison'. This tab itself consists of five tabs. Here you can directly compare the output quantities of the scenarios. Click, for example, on tab **TM**:

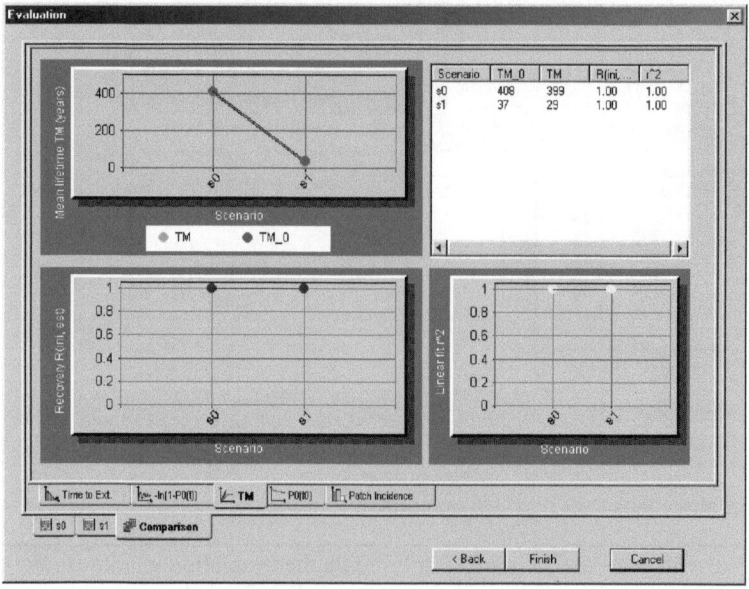

There is one additional feature: you can convert the mean time to extinction to the extinction risk for a certain time horizon, $P_0(t)$. For risk assessment, it is crucial to specify the time horizon of the risk assessment:

4. Choose tab **$P_0(t_0)$**. t0 is the length of the time horizon.
5. In the box 'Enter t0', enter 100 for a time horizon of 100 years.
6. Click **Apply**:

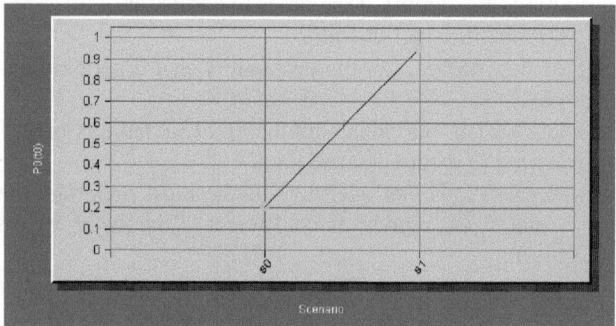

You can see that in the first scenario the risk that the metapopulation will die out within the next 100 years is about 0.2 or 20%, i.e. rather high. In the second scenario, extinction within 100 years is almost certain (about 95%).

Note: This result shows that you have to be careful when interpreting the quantity mean time to extinction, T_m. A T_m of 400 years may sound long, and the difference between 400 or 30 years may sound big, but to have an extinction risk of not more than 1% in 100 years, you would need a T_m of more than 10,000 years (see Chap. 13)!

We can summarize the result of our first experiment: the extinction risk of our model metapopulation is unacceptably high in both scenarios, but a loss of patch P4 would lead to almost certain extinction very soon.

7. To print this graph, or to copy it to the clipboard (and then paste it into another program), click the right mouse button and choose **Print Graph** or **Graph to Clipboard**.
8. Click **Finish**.

To save the evaluation on your hard disk, select **File|Save** in the menu.

5.5 Generating a Report

To document your work done so far on paper, you can automatically generate a report either of the whole experiment or of selected scenarios. A report is a document HTML format, and you will need a WWW browser on your computer in order to be able to generate a report.

1. Choose **Experiment|Report**.
2. Select an experiment. Since there is only one experiment in your project so far, simply click **Next**.
3. Click **Change File** if you want to change the directory where the report is to be saved, and the name of the report file.
4. If you only want to document the comparative results of the experiment but not the individual scenarios, click 'Do not include scenario reports'.
5. Click **Finish**.

Now the report is created. Use a WWW browser to open the report file.

5.6 Where Next?

In this Guided Tour you have learned the most important features of META-X. This – in combination with what you learned in the Chapter 4 – is all you need to work with META-X. If you are not familiar with PVA and ecological models and so did not fully understand what you – or META-X – did during the Guided Tour, try reading Part II of the META-X book.

The other chapters of Part I of this book introduce the features of META-X in more detail and also present some additional features which, although they do not provide any new functions, are very convenient (automatic variation of parameters, landscape editor).

Part II contains detailed information on PVA, metapopulations and the META-X model, helping you to perform your own PVA even without continuous support by modelling and PVA experts.

6 The Project Tree

Overview

The Project Tree in the left half of the META-X screen is designed to organize your projects:

- It hierarchically lists the experiments of a project, the scenarios of an experiment and the parameters and evaluations of experiments and scenarios.
- The most important procedures of META-X (scenario, experiment and simulation wizard) can be started directly from the Project Tree.
- Scenarios may be selected to be displayed in the Landscape Editor.
- You can copy (or cut) and paste existing experiments and scenarios.
- Experiments and scenarios can be deleted or renamed.
- You can modify the sequence of scenarios in the experiment.

6.1 The Elements of the Project Tree

A typical Project Tree is shown on the next page. Note that you can shift the boundary between the windows of the Project Tree and the Landscape Editor by placing the mouse cursor over this boundary, holding down the left mouse key, and dragging the boundary to a new position.

The best way to learn how to work with the Project Tree is the exploratory way: double-click (or click the right mouse key) on all elements of the Project Tree, see what happens and try the options which show up in the context menus.

The Root of the Project Tree

The root of the Project Tree contains the project's name ('Untitled' in the above screen-shot).

To rename the project, select the name by clicking with the mouse and, after a short pause, click again – the name can now be edited.

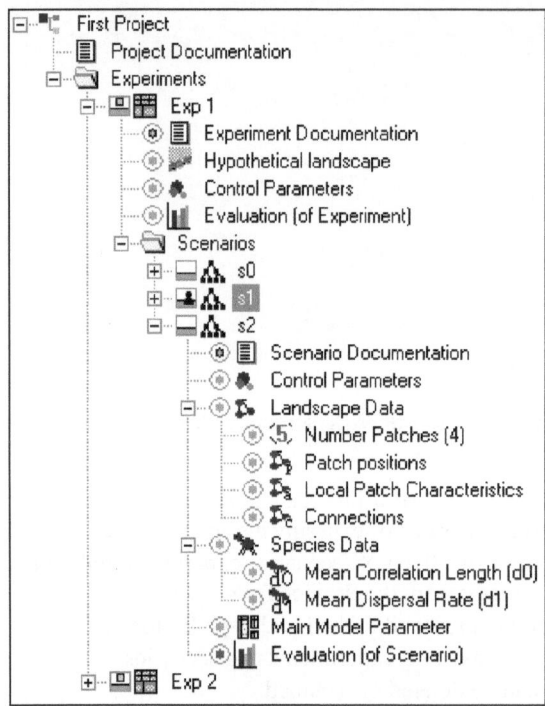

Note: Clicking on the box with the minus or plus symbol will pack or unpack respectively the whole Project Tree; alternatively, double-click the project's name. This is also true for the branches Experiments, the name of experiments or scenarios, respectively, and for Landscape Data and Species Data.

The first element of the Project Tree is the project's documentation, which you can display with a double-click and then edit.

The Experiments Branch

The Experiments branch has a sub-branch for each experiment of the project. The branch of each experiment is organized as follows:

The colour of the diskette-like symbol to the left of the experiment's name indicates whether all scenarios of the experiment are completely specified such that they may be simulated. Green indicates complete, red incomplete. The same colour code is used in all other elements of the tree which may be complete or incomplete.

A double-click on one of the first three sub-branches of the experiment (or choosing **Properties** from the context menu of these three items) opens an overview of the experiment's documentation and parameters:

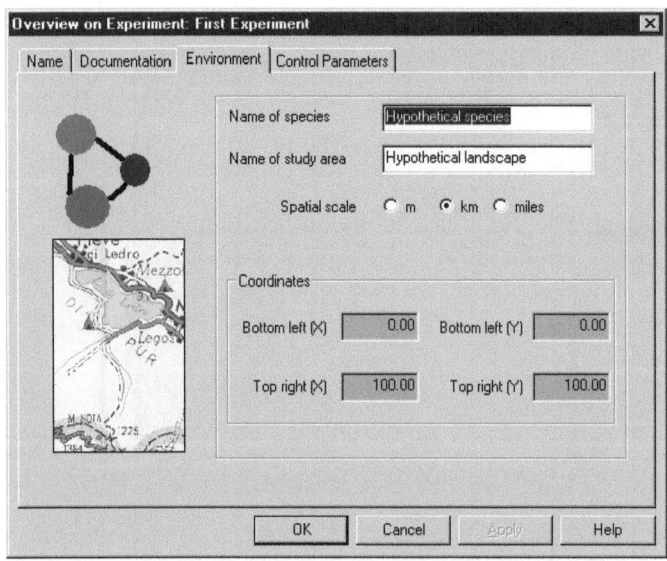

Note that – unless you activated 'Advanced Mode' (see Chap. 12) – you will not be able to change parameters but have to use the wizards; the only exception is the control parameters of an experiment.

A double-click on the fourth sub-branch (**Evaluation (of Experiment)**) opens the evaluation of the experiment. If the experiment has not yet been evaluated, the dot at the beginning of this branch will be red.

The Scenarios Branch

The Scenarios branch has a sub-branch for each scenario of the experiment. The elements of a scenario in the Project Tree are organized as follows:

The colour of the diskette-like symbol to the left of the scenario's name indicates whether the scenario is completely specified so that it may be simulated. Green indicates complete, Red incomplete. If the symbol contains a small, black icon (symbolizing a 'user'), some parameters of this scenarios are 'user-defined', i.e. have not been calculated from the META-X submodels but have been specified or imported by you.

The scenario branch unfolds into six items, of which two – 'Landscape Data' and 'Species Data' – contain sub-branches of their own. If you also expand these sub-branches, there are in total ten items. Apart from the last item, which is the evaluation of the scenario, a double-click on the item (or choosing **Properties** in the context menu of these items) will open a window with nine tagged panels which gives you an overview of the model and control parameters:

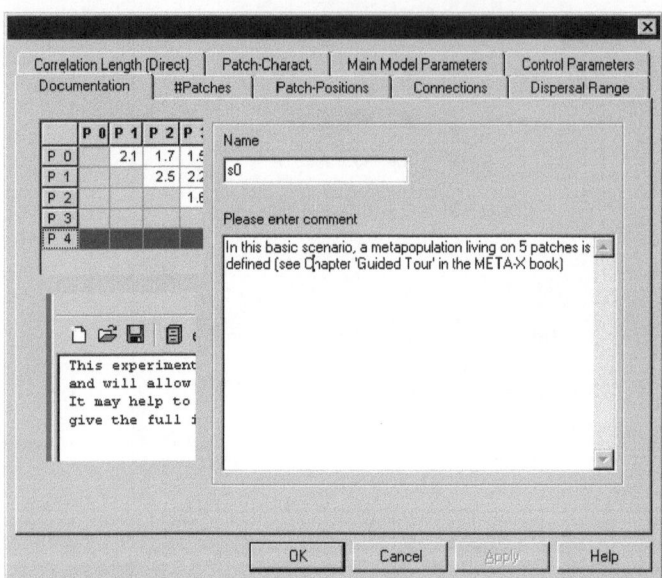

Click on one of the tags to see the corresponding page. If you click on one of the tags in the upper row, the upper and lower layer of tagged panels will be switched.

Note that – unless you activated 'Advanced Mode' (Chap. 12) – you will not be able to change parameters but have to use the wizards; the only exception is the control parameters.

A double-click on the **Evaluation (of Scenario)** item opens the scenario's evaluation. If the scenario has not yet been evaluated, the dot at the beginning of this branch will be red.

6.2 Working with the Project Tree

Using the context menus of the Project Tree, which are activated by a right mouse-click while the mouse cursor points at a certain item of the tree, you can perform most of the procedures of META-X.

Working with Experiments

For experiments, the following procedures are available:

Creating a new experiment

1. Place the cursor over the identifier **Experiments** and click the right mouse key.
2. In the context menu that shows up, choose **New Experiment** and follow the Experiment Wizard.

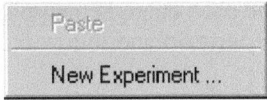

Copy (or cut) and paste experiments

1. Place the cursor over the name of the experiment you want to copy and paste and click the right mouse key.
2. In the context menu that pops up, choose **Copy** or, if you want to cut out the experiment from this project and paste it into another experiment, choose **Cut**.

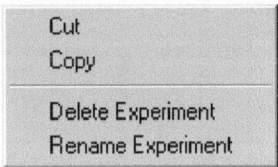

3. Place the cursor over **Experiments** and click the right mouse key.
4. Choose **Paste** in the context menu.

Note: You can paste the experiment into this project or any other project.

Deleting and renaming experiments

Activate the context menu by a right mouse-click over the experiment's name, and choose **Delete** or **Rename**.

Modifying experiments

1. Activate the context menu which is linked to **Experiment Documentation** (and the following two items of the tree).

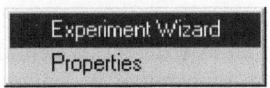

2. Choose Experiment wizard.

Alternatively, choose **Experiment Wizard** in the context menu linked to the items of a scenario.

Working with Scenarios

For scenarios, the following procedures are available:

Creating a new scenario

1. Place the cursor over the identifier **Scenarios** and click the right mouse key.
2. In the context menu that appears, choose **New Scenario** and follow the Scenario Wizard.

Copy (or cut) and paste scenarios

1. Place the cursor over the name of the scenario you want to copy and paste and click the right mouse key.
2. In the context menu that pops up, choose **Copy** or, if you want to cut out the scenario from this experiment and paste it into another experiment, **Cut**.

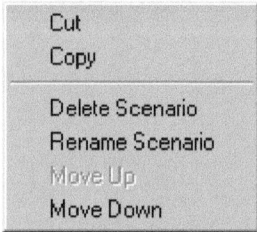

3. Place the cursor over **Scenarios** and click the right mouse key.
4. Choose **Paste** in the context menu.

Note: You can paste the scenario into this experiment, or into any other experiment, but in the latter case make sure that the scenario is based on the same environment, i.e. coordinate system, scale and species.

Deleting and renaming scenarios

Activate the context menu by a right mouse-click over the experiment's name, and choose **Delete** or **Rename**.

Moving scenarios up or down

If you want to change the sequence of scenarios in an experiment (which affects the order of scenarios in the tables and graphs of the comparative evaluation), activate the context menu linked to a scenario's name and choose **Move Up** or **Move Down**. The scenario will be moved up or down by one position in the Project Tree.

Modifying scenarios

1. Activate the context menu which is linked to **Scenario Documentation** (and the following items of the scenario's branch, except 'Evaluation (of Scenario)').

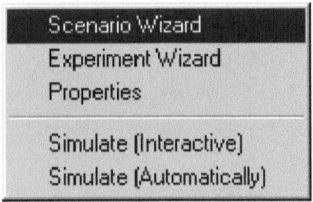

2. Choose **Scenario Wizard** (or **Experiment Wizard** if you want to use the options which are available in the Experiment Wizard).

Alternatively, double-click the scenario's name to load the scenario into the Landscape Editor and than modify the scenario with the Landscape Editor (see Chap. 9).

Starting the simulation of a scenario

Activate the context menu which is linked to **Scenario Documentation** (and the following items of the scenario's branch, except 'Evaluation (of Scenario)'), and choose **Simulate (Interactive)** or **Simulate (Automatically)**.

7 Scenarios and Experiments

Overview

You will spend most of your time working with META-X specifying scenarios and experiments, i.e. with translating your real-world or theoretical problems into parameter sets for the generic metapopulation model which is implemented in META-X. Most elements of how to specify scenarios and experiments are described in the Guided Tour and will not be repeated here in detail; in particular, the screenshots are not repeated. Instead, this chapter gives an overview of:

- The structure and elements of the Experiment Wizard.
- The structure and elements of the Scenario Wizard.
- The specification and purpose of 'homogeneous' parameters.
- The hierarchy of model parameters in META-X which allows you to create 'user-defined' scenarios.

7.1 The Experiment Wizard

The Experiment Wizard is a sequence of input screens which prompts for a series of parameters and, in particular, invokes the Scenario Wizard. The Experiment Wizard allows you to specify new or to modify existing experiments and is started with:

- **Experiment|New** or **Experiment|Modify** in the menu, or
- **New Experiment ...** in the context menu which appears if you right-click the 'Experiments' item in the Project Tree, or
- **Experiment Wizard** in the context menu linked to the item 'Experiment Documentation' (or the following two items) of a certain experiment, or
- **Experiment Wizard** in the context menu linked to the tag 'Scenario Documentation' (or the following items) of a certain scenario.

No matter which of these options you choose, first the window 'Name of Experiment' will appear, and then the control window of the Experiment Wizard (called 'Experiment Wizard'):

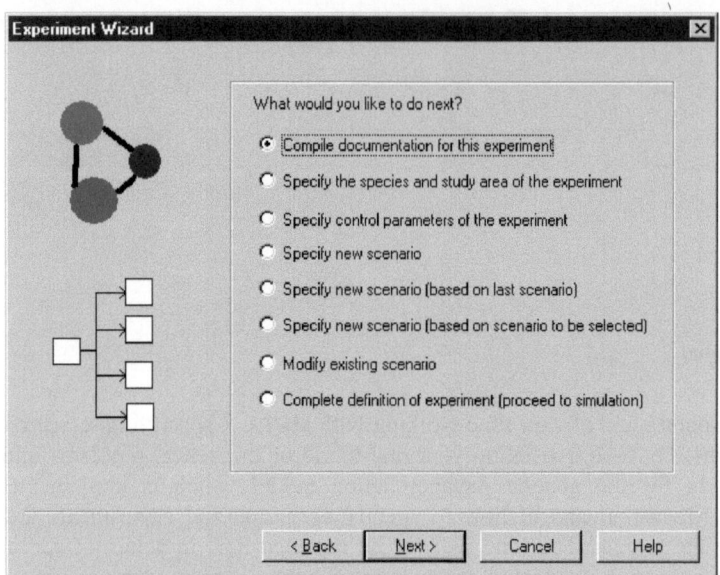

If you are specifying a new experiment, the choices 'Specify new scenario(based on last scenario)', 'Specify new scenario (based on scenario to be selected)' and 'Modify existing scenario' will not be available.

You can choose any of the available choices in any order you want. This means that you will not have to completely specify one experiment (or scenario) in one stroke, but you can leave – if necessary – some of the model parameters unspecified. In this case, when you return to the main program, the experiment will be indicated in the Project Tree as 'incomplete' by the red colour of the symbol besides the experiment's name.

In the following, the choices of the control unit are listed and described.

Compile Documentation

The fields 'Author' and 'Date' document who created the experiment, and when. With the field 'Please enter description' we strongly suggest that you give a complete and self-explanatory description of the experiment: what is the purpose of the experiment, what scenarios are in the experiment, what is the difference between the scenarios, how the parameters were determined, what earlier experiments are related to this experiment, etc. As with any other experiment in the laboratory or in the field, the (computer) experiments of META-X are of little use if not carefully documented.

Specify the Species and the Study Area

Here, enter the 'Name of the species' your experiment is about (if it is a real species and not a theoretical experiment), the 'Name of the study area' and the 'Spatial scale' or unit you will use to specify locations and distances (**km** is default, but you can also use **m** or **miles**). In the 'Coordinates' fields, enter the x and y coordinates of the bottom left and upper right corners of a rectangular section of the real landscape. The coordinates of the bottom left corner are usually (0, 0).

Specify Control Parameters

The control parameters control how often ('Number of runs') and for how many years at most ('Time horizon') the simulation of a scenario will be run.

Specify New Scenario (Based on ...)

There are three alternatives to invoke the Scenario Wizard, i.e. the sequence of screens which prompt you to specify a full set of model parameters (the Scenario Wizard is explained in the next section):

1. 'Specify new scenario' means defining a new scenario from scratch; all input fields are blank and have to be filled.
2. 'Specify new scenario (based on last scenario)' means that all input fields of the Scenario Wizard are filled in with the values of the last scenario in the list of scenarios of this experiment (see Project Tree). This choice is particularly convenient if you – as will often be the case – want to create a series of scenarios which differ by only one or a few aspects from each other, e.g. a certain patch deleted or connection cut, or the extinction rate on a certain patch significantly increased. You will thus only have to click **Continue** until you reach the window with the parameters to be changed and then to continue to the end of the Scenario Wizard.
3. 'Specify new scenario (based on scenario to be selected)'. This is very much the same as the option 'based on last scenario', except that you have to select one of the existing scenarios of the experiment as the basis for the creation of a new one.

Modify Existing Scenario

This is the right choice if you simply want to modify an existing scenario. You will be prompted to select the scenario you want to modify. Then click through the sequence of windows of the Scenario Wizard and make your changes. Do not forget to document your changes in the Scenario Documentation. Note that by using the **Back** and **Continue** buttons you can move back and forth in the Scenario

Wizard. Be careful with the **Cancel** button because it would return you to the main program, losing all your work in this session with the Experiment Wizard.

Complete Definition

If you choose this option, you can in the next window, 'End of Experiment', decide whether you want to proceed to the Simulation Wizard (Chap. 8) or to return to the main program. If you want to make sure that your session is saved to disk, choose 'Return to main program' and then in the window 'What next?' click **Finish**.

7.2 The Scenario Wizard

The Scenario Wizard, i.e. the sequence of screens which prompts you to specify a full set of model parameters, may be invoked in different ways:

- **Scenario|New** or **Scenario|Modify** in the menu.
- **New Scenario ...** in the context menu which shows up if you right-click the 'Scenarios' item in the Project Tree, or
- **Scenario Wizard** in the context menu linked to the item 'Scenario Documentation' of a certain scenario.

The elements of the Scenario Wizard are listed in the following. For further information see:

- Chapter 14 for a detailed explanation of the model parameters to be specified with the Scenario Wizard,
- Chapter 11 for an explanation of the **Read from File** buttons,
- The section below, 'Homogeneous Parameters', for an explanation of the **Set** buttons, and
- The section below, 'User-Defined Scenarios...', for an explanation of where, why and how you may overwrite model parameters that have been calculated by the submodels implemented in META-X.

Documentation of the Scenario

The name of the scenario is automatically constructed from the name of the experiment and the number of the scenario in this experiment, for example 'Habitat<5>'. You may enter a different name, but observe that in the evaluation of experiments, the figures which present the results comparatively will display only the first few characters of the scenario's name. Therefore, short names such as 's1', 's2' etc. are preferable.

For the field 'Please enter comment' the same holds as for the documentation of experiments: we strongly suggest that you give a complete and self-explanatory description of the scenario because scenarios and experiments would not be of very much use if poorly documented.

Number of Patches for a Scenario

For a new scenario, simply enter the number of patches of your system. If you modify a scenario and decrease the number of scenarios, you will have to determine in the window 'Select Patches to be Removed' which patch you want to remove:

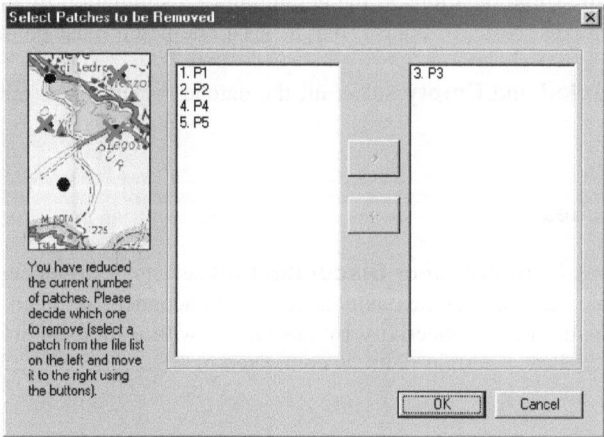

In the left window, click the name of the patch you want to remove, then click on the arrow pointing to the right. If you erroneously remove the wrong patch, you can restore it using the left arrow. Note that if you proceed with the Scenario Wizard, the removed patch will be removed permanently (unless you click **Cancel** somewhere in the Scenario Wizard, returning you to the main program and leaving the original scenario unchanged).

Position of Patches

Enter the X and Y coordinates of all patches: in the entry of the X coordinate of the first patch, select the question mark and type in the value of the X coordinate, for example '25'; then type the tab key on the keyboard and specify the next coordinate, and so on. If you enter a value which goes beyond the border of the coordinate system of the study area, the colour of the corresponding field in the coordinate matrix is red; correct values are displayed in green.

Clicking on the button **Update distances** will calculate the distances between all pairs of patches. If you don't click this button, you will be asked to make up for this.

Patch Characteristics

Enter for each patch the rate of extinction ('Ext. Rate') of a subpopulation living on this patch, the annual number of 'Emigrants per Year' which leave the patch and possibly reach and colonize other patches, and the number of 'Needed Immigrants', i.e. the number of immigrants which would lead to a 50% chance of a new subpopulation becoming established on this patch. You can also determine which patches are occupied by subpopulations at the beginning of a simulation by clicking the 'Occupancy' boxes. Note, of course, that at least one patch must be occupied!

The buttons **Occupied** and **Empty** select all the patches occupied or empty, respectively.

Connection of Patches

Clicking on the buttons **Connect all** or **Disconnect all** selects or deselects, respectively, all possible pairs of patches as connected or unconnected. Then you can define a certain pattern of connectivity by marking – with a mouse-click on the corresponding checkbox – certain pairs as connected or disconnected, respectively.

Correlation Length

The window 'Correlation length' offers two different ways of specifying the correlation length (the model parameter that determines the correlation of extinction events between pairs of patches):

1. **Enter the mean correlation length (direct method)**: you will have to enter the mean correlation length directly.
2. **Select two references ...(indirect method):)**: for a selected pair of patches, you will have to enter the correlation of extinction events, i.e. the probability that extinctions occur simultaneously on these patches.

In most cases you will use the direct method, but sometimes you might have specific information on the correlation of extinction events between a certain pair of patches.

Correlation Length (Direct)

Enter the correlation length. The scale (m, km, or miles) will be as chosen in the experiment's 'Environment'. **Reset matrix** will calculate the correlation matrix, i.e. the correlation of extinction events between all pairs of patches. This calculation is based on the META-X submodel of correlations (see Chap. 14).

User-defined: You can overwrite all or selected elements of the correlation matrix to take into account certain knowledge about the landscapes structure, for example that certain patches are on south-exposed hill-slopes and therefore, irrespective of their distance, have a very high correlation of extinction events, whereas south- and north-exposed patches may have a very low correlation even though they are very close to each other.

Correlation Length (Indirect)

Choose a pair of patches in the corresponding field:

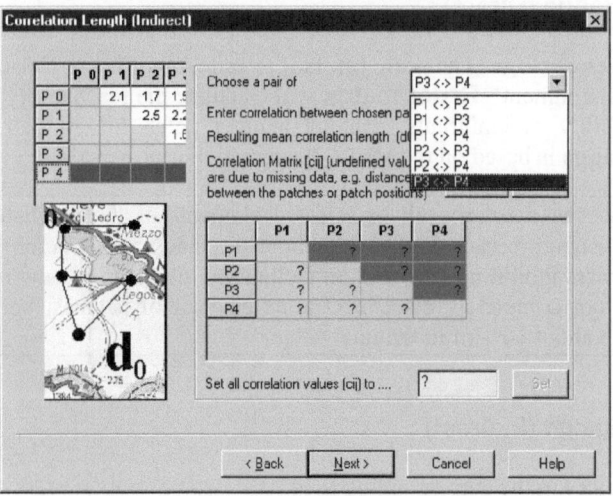

Then enter the correlation of extinction events between these two patches (note that we consider positive correlations only; therefore the degrees of correlation can take values from the interval $[0,1)$). The resulting mean correlation length is calculated automatically and is displayed in the field 'Resulting mean correlation length (d0)'. **Reset matrix** then calculates the correlation matrix. You will be warned that **Reset matrix** will overwrite all user-defined input in the correlation matrix with results obtained from the META-X standard submodel of correlation. To make user-defined changes, see above (section 'Correlation Length (Direct)' and the section 'User-defined scenarios' below).

Mean Dispersal Range

As with correlations, there are two different ways to specify 'mean dispersal range', which is the model parameter describing dispersal distances and, in turn, the reachability between all pairs of patches:

1. **Enter the mean dispersal range directly** You will have to specify the model parameter 'mean dispersal rate' directly.
2. **Select two (connected) reference patches...**: The model parameter 'mean dispersal rate' is calculated from the reachability between a selected pair of patches. 'Reachability' is the probability that an emigrant from a certain patch will reach another certain patch.

In most cases, you will use the direct method, but in some cases you might have specific information on the reachability between a certain pair of patches which then, assuming a more or less homogeneous landscape, can be used to calculate the mean dispersal range for the entire landscape.

Mean Dispersal Range (Direct) :

Enter the mean dispersal range. The scale (m, km, or miles) will be as chosen in the experiments 'Environment'. **Reset matrix** will calculate the reachability matrix, i.e. the probability of emigrants reaching another patch for all pairs of patches. This calculation is based on the META-X submodel of dispersal.

User-defined: You can overwrite all or selected elements of the reachability matrix to take into account certain knowledge about the landscape structure, for example certain barriers inhibiting or corridors facilitating dispersal. A new road, for example, might not completely disconnect a certain pair of patches, but decrease reachability to about 10% of its original value.

Mean Dispersal Range (Indirect) :

Choose a pair of patches in the corresponding field:

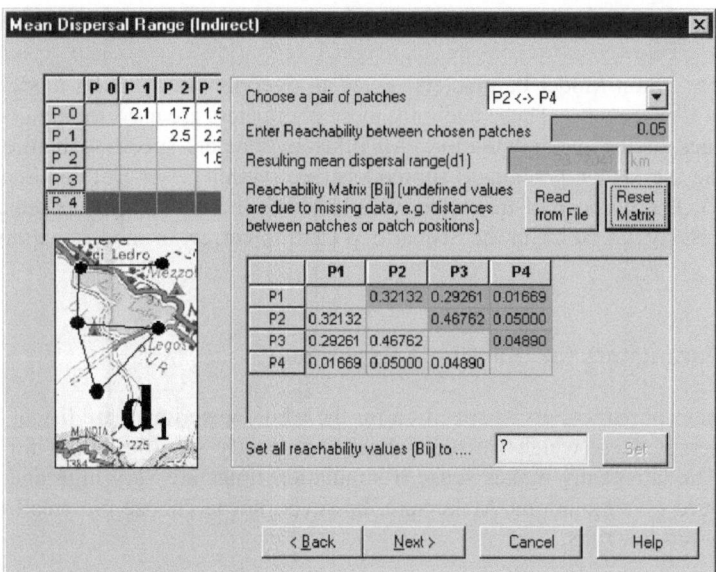

Now enter the reachability for this pair of patches. Note that the reachability is a probability and thus has to be in the interval [0,1). The resulting mean dispersal d_1 range is calculated and displayed automatically. **Reset matrix** will calculate the reachability matrix. You will be warned that **Reset matrix** will overwrite all user-defined input in the reachability matrix.

Rates of Colonization

This window shows the colonization matrix, i.e. the rate of colonization between pairs of patches. Of course, this rate is zero for pairs of patches which are not connected. Note that in META-X, 'colonization rate' $b_{i,j}$ means the rate by which emigrants from patch i reach an empty patch j and then establish a population on this patch. $b_{i,j}$ is thus calculated from the reachability matrix and the patch parameters 'Emigrants' and 'Immigrants needed'. Thus, the calculation of the **colonization** matrix is either based on the META-X submodels of emigration, dispersal, and colonization, or on your user-defined input into the reachability matrix.

User-defined: You can overwrite all or selected elements of the colonization matrix (you can even completely leave out the reachability matrix and just enter the colonization matrix). This is necessary if you do not want to use the standard submodels of META-X, but your own, external submodel on colonization. In this case, it would be most convenient to export the colonization matrix from your external model or program, respectively, and to import it using **Read from file** (see Chap. 11 and the section 'User-Defined Scenarios' below).

Main Model Parameters

The window 'Main Model Parameters' gives an overview of all main model parameters of this scenario. These and only these parameters are used for simulating the dynamics of the metapopulation. All other parameters specified before are only used in the META-X standard submodels (see section 'User-Defined Scenarios' below). The purpose of this window is to check the main model parameters and, if necessary, go **Back** in the Scenario Wizard to correct or modify parameter values.

Control Parameters for the Whole Experiment

Here, the control parameters as specified for the whole experiment are listed. You can use these values (which we recommend) or specify specific values for this scenario. The latter only makes sense if simulation times are very high and you want to speed up simulations. Make sure, however, not to choose too small time horizons or too few runs.

End of the Scenario Specification

Now you are at the end of the Scenario Wizard. If both the main model and the control parameters are completely and correctly specified (i.e. within their allowed intervals), the specification or modification of the scenario is finished and the scenario will be marked as 'complete' in the Project Tree (green symbol beside the scenario's name), otherwise as 'incomplete' (red symbol). If you click **Finish** now, you will return to the main program or the Experiment Wizard, depending on how you invoked the Scenario Wizard. If the main model parameters of this scenario do not tally with the standard submodels of META-X, the symbol besides the scenario's name in the Project Tree will be marked as 'user-defined' by a user icon in the scenarios symbol:

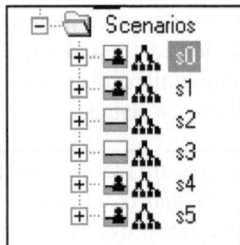

7.3 Homogeneous Parameters

The metapopulation model of META-X is spatially explicit: the patches have an explicit location and the different distances between patches matter with respect to both correlation and colonization. However, the original metapopulation model from 1969 is spatially implicit: only the occupied fraction of an – in principle – infinitely large number of identical patches is considered; distances and locations do not occur. For modelling real populations, Levins' model certainly is oversimplified, but it is still useful for demonstrating the idea of the metapopulation concept and thus for teaching and understanding the general aspects of metapopulation dynamics.

To make a scenario of META-X spatially implicit and patches identical, all model parameters could be specified accordingly by hand, e.g. making all local extinction rates the same. For the sake of convenience, however, in the following steps of the Scenario Wizard you can use the **Set** button to make all the parameters the same:

Name of window in Scenario Wizard	Input field
Position of Patches	'Set all distances (dij) to ...'
Patch characteristics	'Set rate of extinction to ...'
	'Set mean number of emigrants to ..'
	'Set mean number of immigrants to ...'
Correlation Length (Direct/Indirect)	'Set all correlation values (cij) to ...'
Mean Dispersal Range (Direct/Indirect)	'Set all reachability values (Bij) to ...'
Rate of Colonization	'Set all rates of colonization (bij) to...'

In each of these cases, enter a value in the field and click the **Set** button. Note that of course scenarios where all distances, correlations, reachabilities, or rates of colonization are equal are theoretical scenarios which cannot be displayed graphically. Therefore, the Landscape Editor will either not display the patches (if you did not enter patch positions at all) or, if you entered patch positions but then chose a homogeneous distance, will announce in the Status field, 'Patch positions are not consistent with distances'.

If you use homogeneous parameters, you will – as you proceed through the Scenario Wizard – be warned that the parameters no longer tally with the standard submodels of META-X. Make sure to choose not to update the matrices automatically, because this would overwrite the homogeneous parameters!

7.4 User-Defined Scenarios and the Parameter Hierarchy of META-X

One decisive concept of META-X is the hierarchy of model parameters. To simulate metapopulation dynamics, only the main model parameters are needed:

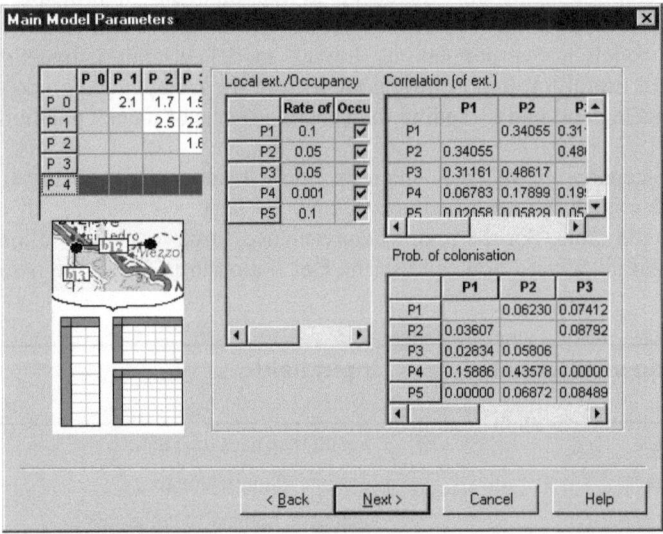

1. Rate of extinction of each patch (or, to be precise, of subpopulations which live on this patch), ν_i, i.e. the inverse of the mean time to extinction of the patch's population.
2. Correlation of extinction events between pairs of patches, c_{ij}, i.e. probability that subpopulations on patches i and j go extinct in the same year.
3. Rate of colonization between pairs of patches, b_{ij}, i.e. probability per year that emigrants from patch i reach the empty patch j and establish a new subpopulation there.

Submodels

All other parameters required by the Scenario Wizard are needed for the standard submodels of META-X, i.e. for models which are used to calculate the main model parameters. Although these submodels are very general, they will not be applicable to all ecological situations. For example, in the META-X dispersal model, all emigrants of a patch are equally divided among the number of connections to other patches. However, this might be more or less unrealistic in many situations, and therefore you would have to develop an external submodel of your own and enter the results of these submodels into META-X (using, if possible, the **Read from file** option).

The Standard Submodels of META-X

1. Rate of extinction on patch i, v_i : META-X provides no submodel for this parameter because first, META-X does not simulate local population dynamics (instead, only the presence or absence of a subpopulation is considered), and second, because we think that a generic submodel of all possible species is not feasible. You thus have to use an external submodel here (e.g. another generic PVA software, or the submodels listed in Chap. 15, or your own simulation programs).
2. Correlation of extinction events between pairs of patches, c_{ij}: patch distances and the mean correlation length, c_0, are used to calculate c_{ij} (see Chap. 14).
3. Rate of colonization between pairs of patches, b_{ij} : patch distances and the parameters 'mean dispersal range', 'number of emigrants per year' and 'number of immigrants needed' are used to calculated b_{ij} (see Chap. 14).

User-Defined

'User-defined' means by definition that entries in the correlation, reachability, or colonization matrix differ from those that would have been calculated using the submodels of META-X. In other words, you overwrote entries in the correlation, reachability, or colonization matrix. You may modify only one or a few matrix elements, for example if you want to take into account specific landscape features, such as barriers. Alternatively, you may completely determine the matrix values by an external submodel (which could then easily be imported using **Read from file**; for the format of the import file, see Chap. 11) or using the **Set** button (homogeneous parameters, see above).

Parameter Hierarchy

You can enter your user-defined parameters at different levels of the parameter hierarchy:

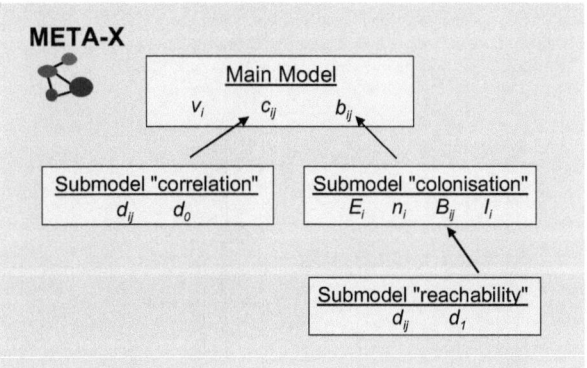

On top are the main model parameters. It is possible to specify only these parameters, i.e. local extinction rates, correlation and colonization rates, and to leave all other parameters occurring in the Scenario Wizard blank – the scenario will nevertheless be marked as complete, and user-defined.

Update Matrix: Warning Message

As you go back and forth in the Scenario Wizard, META-X checks if the entries in the parameter matrices tally with the META-X submodels. If this is not the case, you will be warned that this is so:

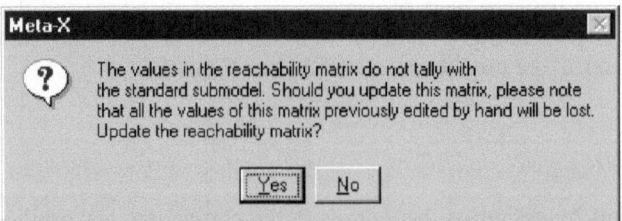

The purpose of this message is to make sure that you are constantly aware of whether you use edited or imported parameter matrices, or matrices generated by the standard submodels.

Now you are asked if you want to update the matrix. Choose **No** if you want to keep your non-standard matrix, or **Yes** if you want to replace it with the matrix generated by the standard submodel.

Note: You are very likely to encounter this warning message and dialog very often. In most cases, you will choose **No**, but make sure you always give it a second's careful thought before you click **Yes** or **No**.

8 Simulation and Evaluation

Overview

As soon as you have specified scenarios and experiments, the time has come for the computer to do its job: to simulate metapopulation dynamics for a certain number of years, e.g. 300 years. 'Simulation' means starting with a certain state of the model metapopulation (i.e. a certain pattern of patch occupancy) and then using the model parameters to calculate the state of the metapopulation after one time step. This new state is then used to calculate the state after the next time step, and so on.

However, one single simulation does not tell us much about viability. Due to the stochastic nature of extinction and recolonization, the basic quantities calculated by META-X to assess extinction risk – mean time to extinction and risk of extinction within a certain time interval – are statistical quantities which are gained from many (typically 1,000) simulations.

Therefore, to assess viability or risk of extinction, respectively, META-X will automatically repeat simulations. You may automatically simulate individual scenarios, a set of scenarios within an experiment or whole experiments. The results of the simulations are stored in the evaluations.

There are also interactive simulations which allow simulations to be run interactively. You can thus follow at one time step at a time how patches go extinct and are recolonized depending on the landscape structure and the main model parameters. This enables you to acquire an intuitive feel for and an understanding of the dynamics of the metapopulation because you will 'see' the significance of each patch and each connection between patches for overall persistence.

8.1 Interactive Simulation

To interactively simulate a scenario, first select a scenario:

1. Choose **Scenario|Interactive Simulation** in the menu, and select – in the window 'Select Experiment and Scenario' – an experiment and then a scenario, or

2. In the Project Tree, open **Experiments**, open **Scenarios** and open the scenario that you want to simulate; click on one of the items of the scenarios branch (except 'Evaluation'), activate the context menu with a right mouse-click and choose **Simulate (Interactive)**:

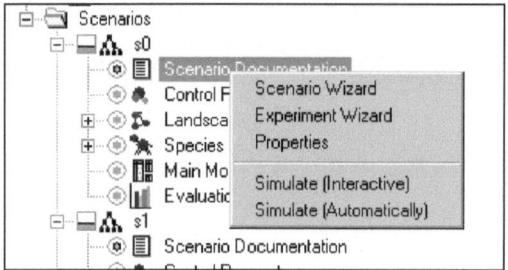

This opens the window of the Interactive Simulator. All features of the interactive simulator are accessible from this control window:

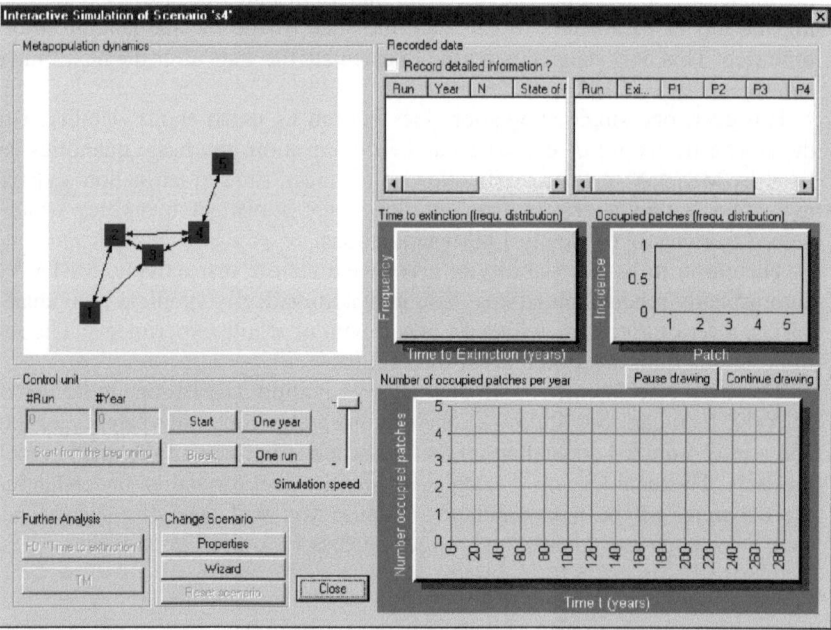

The control window has the following panels (see also corresponding section in the Chap. 4):

Metapopulation Dynamics

Here the landscape defined in the scenario is visualized. Patch icons in blue indicate occupied patches, icons in grey unoccupied patches (i.e. the corresponding subpopulation is extinct). Click repeatedly **One Year** in the **Control unit** and observe how the occupancy pattern changes; or click **One run** to observe the dynamics (if the dynamics is too fast, slow down **Simulation speed** by using the corresponding control).

Recorded Data

The right table shows for each individual simulation the year of extinction (or the year when simulation stopped, i.e. the time horizon), and the incidence of the patches (i.e. the fraction of simulation years where the patch was occupied). If the checkbox 'Record detailed information' is active, the left table shows for each single year the number of simulation run, current year, number of patches occupied in this year, and the occupancy of the entire metapopulation ('1' indicates an occupied patch, '0' an empty one).

Diagrams

The diagrams show:

- 'Time to Extinction (Freq. Distribution)', the frequency distribution (histogram) of the times to extinction.
- 'Occupied Patches (Freq. Distribution)', the incidence of the patches.
- 'Number of Occupied Patches per Year', the time series of the number of occupied patches. **Pause drawing** will considerably speed up simulations.

Control Unit

Here you can control the interactive simulation:

- **One year**: One time step is simulated.
- **One run**: One run is simulated, i.e. until the population is extinct or the time horizon reached.
- **Continue/Break**: The simulation continues automatically or is interrupted.
- **Start from beginning**: The simulation is reset to the initial state and all results obtained so far are withdrawn.
- **Start**: Starts the Simulation.

During the simulation, the actual year and number of simulation (= run) are displayed.

Further Analysis – FD

The purpose of 'Further analysis' is to learn how the mean time to extinction is calculated from the frequency distribution (histogram) of extinction times.

'FD (= frequency distribution) "Time to extinction"' allows the frequency distribution (histogram) of the extinction times recorded so far to be modified. The default class width of the histogram is the length of the time horizon divided by 50. You can modify the years per class of the histogram and the minimum and maximum year of the distribution:

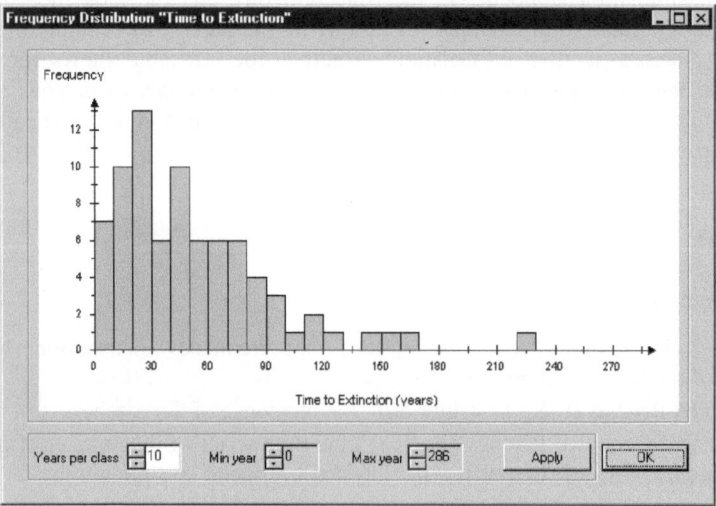

Apply will apply your modifications to the distribution, which then will be used to calculate T_m.

Further Analysis – TM

Once you have simulated a couple of at least, say, 25 runs, and after specifying the properties of the frequency distribution (see above), the analysis **TM** (= mean time to extinction, T_m) demonstrates step by step how the mean time to extinction is calculated from the frequency distribution of extinction times:

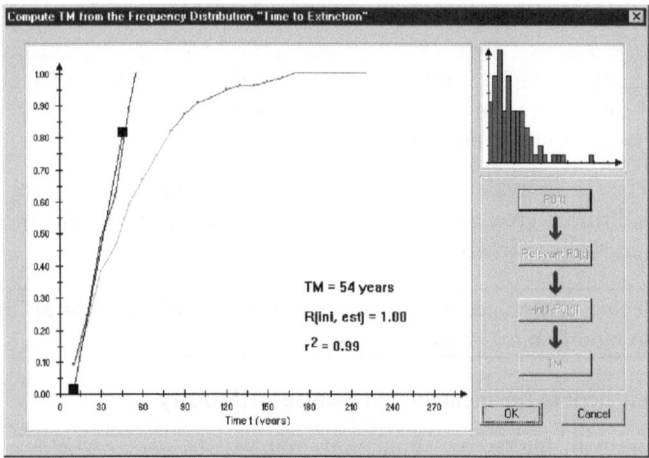

You can again see the histogram of extinction times in the small inlet in the upper right-hand corner. Now click one after the other:

- **P0(t)**: the probability of extinction by time t, $P_0(t)$ vs. time t, determined from the histogram by successively counting the cumulative number of extinctions up to time t and by dividing these cumulative numbers by the total number of runs (red curve).
- **Relevant P0(t)**: Values of $P_0(t)$ larger then 0.97 are ignored because they are strongly influenced by a small number of long-lasting simulation runs (see Chap. 13 for further explanation) (green curve).
- **-ln(1-P0(t))**: the negative natural logarithm of $(1-P_0(t))$ which should – as can be shown theoretically – yield a curve which can be fitted with a straight line (blue curve).
- **TM**: T_m is the inverse of the slope of the (black) regression line.
- **R(ini,est)** is the probability that the metapopulation, which is specified by the initial occupancy pattern, becomes established (see Chap. 13). r^2 is the regression coefficient of the regression line.

Note: You can try and adjust the regression line by hand, i.e. by dragging the tags at the ends of the regression lines.

Observe that T_m and the other quantities determined by META-X are not deterministic quantities but determined from stochastic simulations. Hence do not bother with any small variations in these quantities which may occur – even in an experiment consisting of copies of the same scenario – if in **Tools|Options| System** you deactivate 'Use static seed for simulation'.

Change Scenario

Here you can – assuming you have interrupted the simulation – check and modify control and model parameters and then observe with the Interactive Simulator how metapopulation dynamics and extinction risk change:

- **Properties**: This is equivalent to double-clicking the tag 'Main model parameters' in the Project Tree, i.e. it gives an overview of all model parameters and allows the control parameters to be modified.

Note: If you selected 'Advanced Mode' in the menu (**Tools|Options| System**), you can also modify model parameters here.

- **Wizard**: Is equivalent to the menu option **Scenario|Modify**.

Once you are done with interactive simulation, you will be asked whether you want to save these changes and thereby modify your original scenario.

Note: For a statistically sound assessment of extinction risk, the interactive mode of simulation would be far too slow. This is thus the task of automatic simulations.

8.2 Simulation and Evaluation of Scenarios

In META-X, you will usually simulate and evaluate entire experiments (see the next section); but if, for example, you modified only one scenario of an experiment (which renders the evaluation of this scenario 'undefined'), you don't need to simulate the entire experiment again but can run automatic simulations of individual scenarios.

Automatic Simulation

Choose Automatic Simulation, either from the menu **Scenario| Automatic Simulation** or from the Project Tree by choosing **Automatic Simulation** instead of **Interactive Simulation** (see section 'Interactive Simulation').

You can then decide to record detailed data which you could then export (**File|Export Data**) for further analysis with other programs (see Chap. 11 for a detailed description of these data). Usually, however, you will not record detailed data.

Evaluation

To see the evaluation, i.e. the evaluation of the simulation results, choose **Scenario|Evaluation** from the menu, or double-click the corresponding **Evaluation** tag in the Project Tree. The evaluation window displays:

- The frequency distribution (histogram) of extinction times
- The plot of $P_0(t)$ (the green, 'relevant' part is used to produce the $\ln(1-P_0(t))$ plot) and $-\ln(1-P_0(t))$ versus time t
- The patch incidence, i.e. the probability of each patch being occupied by a sub-population.
- Mean time to extinction (T_m), the probability of reaching the established state (R(ini,est)), and the regression coefficient r^2 of the linear fit to the $-\ln(1-P_0(t))$ plot:

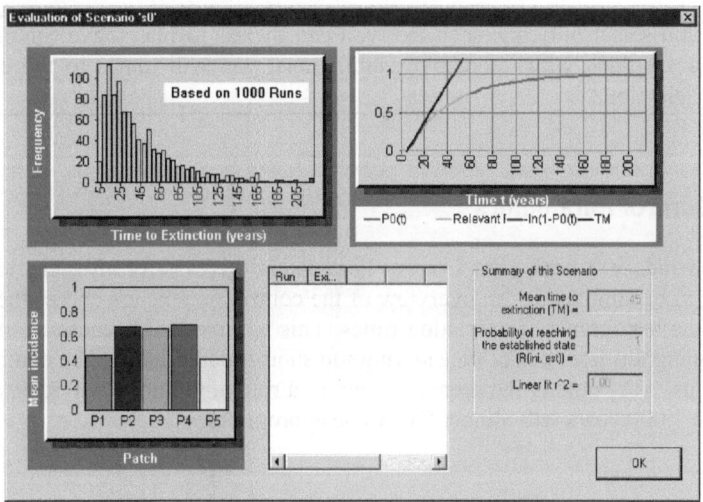

8.3 Simulation and Evaluation of Experiments

One basic concept of META-X is that scenarios which need to be evaluated comparatively are grouped in – comparative – experiments. Therefore, simulating experiments allows the joint simulation of all scenarios of the experiment and, in turn, comparative results to be obtained which display, for example, the extinction risk of all scenarios in one graph.

Select Experiments and Scenarios

Choose **Experiment|Simulate** from the menu and select the experiment and the scenarios of this experiment that you want to simulate. To mark more than one scenario for simulation, press the Shift or CTRL key and click on the scenarios in the list:

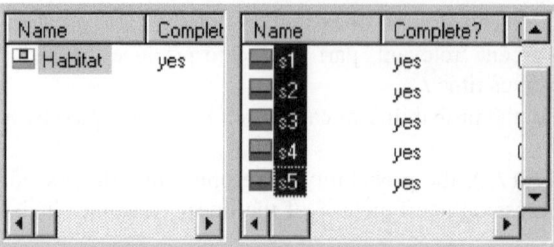

Note: If you select only one scenario, you can bypass to Interactive Simulation. When you are done with interactive simulations, you will return to the current window 'Selection...'.

Check Control Parameters

The next window 'Check your Control Parameters' gives – for all scenarios that are going to be simulated – an overview of the control parameters and of the class width of the histograms of extinction times. This is important because of the risk of performing too few runs or considering too short time horizons. The control parameters may vary among the scenarios, but, as a rule of thumb, 1,000 runs, a time horizon of 1,000 years will almost always be appropriate.

Batch Simulation

In the window 'Batch simulation', the simulation is:

- Started by **Start.**
- Cancelled by **Cancel**.

During simulation, by default the progress of the simulation is indicated by the number of the scenario currently simulated and by the number of runs simulated for this scenario. For very slow simulations you might also activate the checkbox **Years** to check the progress of the simulation. However, deactivating both checkboxes **Years** and **Runs** will speed up the simulation somewhat.

What to Evaluate?

The simulation is finished now and the raw results of the simulation are stored temporarily (but not yet permanently on disk). In the window 'What Do You Want To Evaluate?' you determine:

- ***Quantities to be evaluated***: Click the quantities you want to draw from the simulation results. 'Date (per Run)' means that the time to extinction and patch incidence are recorded for each single run.
- ***Experiment & Scenarios***: Select **Whole experiment** if you want to record the comparative evaluation of the whole experiment; if you want to record more detailed information about all or some of the scenarios (see section 'Simulation and Evaluation of Scenarios'), select the corresponding scenarios:

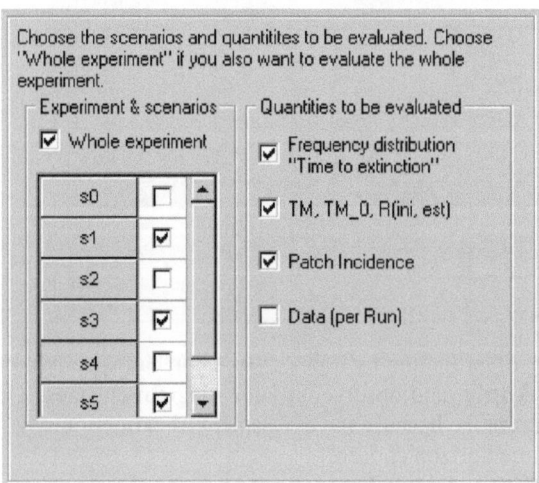

Evaluation

In the window 'Evaluation' the evaluation (i.e. the results of the simulation) are displayed on a couple of tagged panels. The panels of the individual scenarios are the same as those obtained when you check the evaluation of a single scenario (see section 'Simulation and Evaluation of Scenarios').

The panel **Comparison** consists of five (resp. six if you selected 'Data per Run') panels which are selected by clicking on one of the tabs:

- ***TM***: This is the most important panel displaying comparatively for all scenarios of the experiment T_m, $T_{m,0}$, $R(\text{ini, est})$, and r^2 (see Chap. 13).

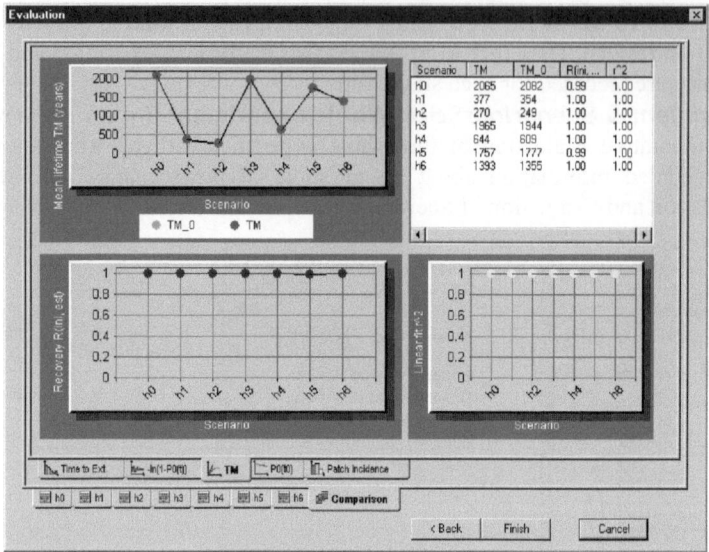

- **P0(t0)**: The probability of extinction by time t_0. Here you can specify any horizon t_0 you want, click **Apply**, and observe how the diagram changes. This option is of particular importance because the assessment of extinction risk necessarily depends on the time horizon chosen.
- **Time to Ext., -ln(1-P0(t)), Patch Incidence**: Here, for better comparison, the corresponding figures of the scenarios are displayed together on one panel.
- **Recorded data:** Each line shows: the number of the run and the year of extinction in that run for each of the scenarios. A '+' indicates a run where the metapopulation survived until the end of the time horizon.

In addition, the evaluations of all scenarios of the experiment that you choose for evaluation are on separate windows (if you chose them for evaluation) which you may select by clicking on the corresponding tag which is labelled with the scenario's name. To save the evaluation of the experiment, click **Finish**!

Once you have simulated an experiment, at any time you can choose **Experiment|Evaluate** from the menu and perform a new evaluation of the simulation results. If in the meantime a scenario has been modified or added, it will be marked as 'undefined' in the evaluation, and so you would have to simulate this scenario or the whole experiment.

Note that clicking on an experiment's **Evaluation** in the Project Tree will display only the comparative evaluation. The evaluations of the individual scenarios can be displayed by double-clicking on the scenario's evaluation branch in the Project Tree.

9 The Landscape Editor

Overview

The right-hand part of the META-X window contains the Landscape Editor which is designed to visualize the landscape of a scenario, to modify existing scenarios via a graphical interface, and to create new scenarios with this graphical interface. Once a scenario has been selected in the Project Tree by a double-click or by choosing **Editor|Load Scenario** in the menu, the Landscape Editor will indicate:

1. Whether the scenario is completely specified or some parameters regarding patches, dispersal range or correlation length are still missing.
2. The position of the patches and if a patch is occupied by a subpopulation.
3. The scales of the scenario, i.e. the basic scale of the map, dispersal range and correlation length.
4. If a so-called 'Local Aspect' is chosen, different aspects of the patches, i.e. the mean time to extinction, number of emigrants produced or number of immigrants needed to establish a new subpopulations (with 50% probability).
5. The connections between patches, i.e. all pairs of patches which can recolonize each other via dispersal.

9.1 Elements of the Landscape Editor

The window of the Landscape Editor contains the following elements:

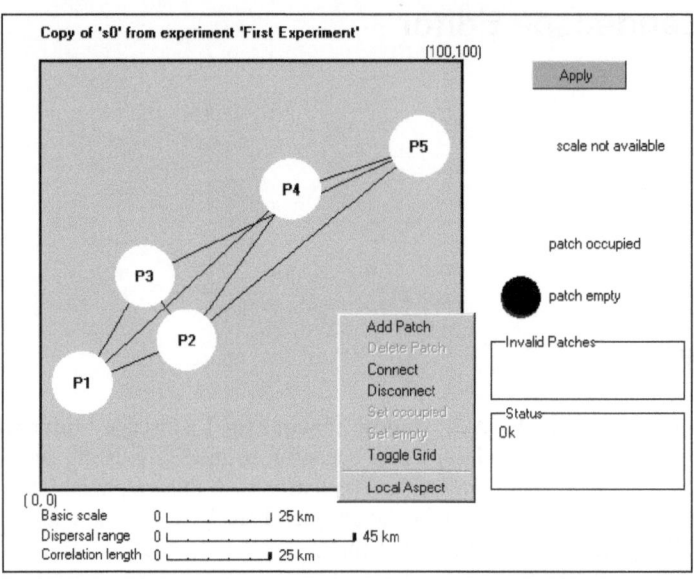

Element	Meaning
Title: 'Copy of ,..'...'	Indicates which scenario is selected. If this title is red, modifications of the scenario produced using the Landscape Editor have not yet been saved. You are thus working with a copy of the scenario until you have saved the changes, not with the scenario itself. This has the advantage that you can reject a change without destroying the original scenario.
Apply button	Saves temporary changes.
Landscape with patches and connections between patches	Shows the map underlying the scenario. The coordinate system of the landscape is marked at the lower left and upper right corner.
Basic Scale	Scale of the map.
Dispersal range	Indicates the model parameter 'mean dispersal range'.
Correlation length	Indicates the model parameter 'mean correlation length'.
Scale not available	Indicates which aspect of the patches is used to scale patch size on the landscape. As long as no aspect is chosen, the label is 'Scale not available'.
Patch occupied/empty	Indicates colours of occupied/empty patches.
Invalid patches	Contains patches which have not yet been assigned a position on the landscape.
Context menu	This menu – which starts with the item 'Add Patch'

	– is the main tool for working with the Landscape Editor. It is activated by a right mouse-click.
Status	Status information about the specification of the scenario.

9.2 Visualization

The Landscape Editor can be used to visualize the landscapes which are defined in the scenarios. There are two ways of loading a scenario into the Landscape Editor:

Loading a scenario into the Landscape Editor using the menu

1. Choose **File|Open** in the menu to choose a project, for example 'First Project.mtx', which was developed during the Guided Tour.
2. Choose **Editor|Load Scenario** in the menu.
3. Select an experiment with a mouse-click, then a scenario.
4. Click **OK** (alternatively, double-click the scenario you want to view or edit).

Loading a scenario into the Landscape Editor using the Project Tree

1. Choose **File|Open** in the menu to choose a project, for example 'First Project.mtx', which was developed during the Guided Tour.
2. In the Project Tree, first select an experiment with a mouse-click, then a scenario. If necessary, expand the experiment or scenario sub-tree by a mouse-click on the corresponding '**+**'- symbols, or by double-clicking on the labels 'Experiments' or 'Scenarios', respectively.
3. Selecting a scenario by a double-click on its label in the Project Tree will load the scenario into the Landscape Editor.
4. To compare several scenarios visually, select them one after the other.

Customizing the Landscape Editor

You can customize the aspect of the Landscape Editor via the **context menu** and via **Tools|Options** in the menu. In the following procedures, we will assume that you have already loaded a scenario to the Landscape Editor.

Changing local aspects with the context menu

1. Place the mouse cursor somewhere within the window of the Landscape Editor.
2. Click the right mouse key.
3. In the context menu which appears, select **Toggle Grid** to toggle a grid in the landscape.

4. In the context menu, select **Local Aspects** to change the aspect of the patches.

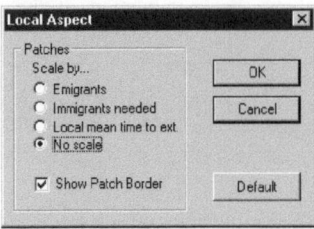

5. In the window 'Local Aspects', select which property of a patch you want to use to scale the size of the patch symbols in the Landscape Editor, i.e. local mean time to extinction, emigrants or immigrants needed. Select 'no scale' if you do not want to scale the patch size.

Note: Once you choose to scale the patch size with one of the patch characteristics, this scale will be indicated in the patch symbol underneath the **Apply** button. If you double-click this patch symbol, the window 'Patch Scaling' opens which allows an absolute scale to be specified instead of one relative to the maximum value of the current scenario. This might be useful if you want to visually compare scenarios with different maximum values of for example local mean time to extinction. For this purpose, choose the same patch scaling for all scenarios.

6. Click 'Show Patch Border' to toggle the display of patch borders.

Note: The colour of the patch border shows whether a patch is completely parameterized. A red border indicates incomplete specification of patch characteristics. Likewise, a red connection indicates that the dispersal range or correlation length still have to be specified.

All the changes you had made here to 'Local Aspects' will also apply when you view other scenarios. Click **Default** if you want to restore the default settings of the 'Local Aspects'.

Further customizing via Tools|Options

1. Choose **Tools|Options** in the menu.
2. In the window 'View Properties', on the tagged panel 'Default Options for View',

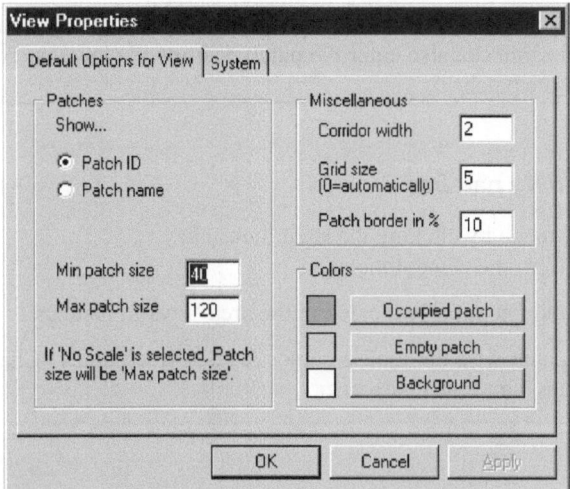

you may:

- Choose whether patches are to be labelled with their identity number (**Patch ID**) or with their name (**Patch name**);
- Adjust the default range of patch sizes which is used to indicate the local aspects of the patches;
- Adjust the width of the corridor lines, the width of the patch borders and the grid size;
- Change the colours of the landscape, i.e. background, occupied and empty patch.

9.3 Modifying Existing Scenarios

You can use the Landscape Editor to change any aspects of a scenario, provided the standard submodels of META-X are used (see Chap. 7):

- *Landscape configuration*: number and position of patches, links between patches.
- *Local parameters*, i.e. patch characteristics.
- *Landscape parameters*: dispersal and correlation length.
- *Control parameters*.

Changing the position of patches

1. Select a patch with a mouse-click.
2. Hold down the left mouse key while it is located over the selected patch and drag the patch to a new position. While doing this, a small window attached to the patch opens and shows the current X and Y coordinates of the patch.

Note: If you find it too laborious to precisely specify patch positions by using this procedure, do not worry: you can also enter the patch position via local parameters (by double-clicking on the patch).

Adding and positioning patches

1. Activate the context menu by clicking the right mouse key.
2. Select **Add Patch** from the context menu.

A new patch is added and temporarily located in the area 'Invalid Patches'.

3. Drag and drop this patch with the mouse cursor (as described under 'Changing the position of patches') into the landscape.

Deleting patches

1. Place the mouse cursor on the patch you want to delete.
2. Open the context menu with a right mouse-click.
3. Select **Delete Patch**.

Adding links between patches

1. Open the context menu with a right mouse-click and select **Connect**.
2. The mouse cursor turns into a cross with a capital S as an index (indicating: Select). Place this cursor on one of the two patches you want to connect.
3. Hold down the mouse key and move the cursor to the other patch.
4. Release the mouse key.

Note: A connection indicates the pairs of patches between which recolonization is possible. The lines representing the connections do not indicate the position of corridors between patches.

Deleting links between patches

1. Open the context menu with a right mouse-click and select **Disconnect**.
2. The mouse cursor turns into a cross with a capital S as an index (indicating: Select). Place this cursor on one of the two patches you want to disconnect.
3. Hold down the mouse key and move the cursor to the other patch.
4. Release the mouse key.

Changing local parameters

1. Double-click on the patch whose parameters you want to modify or specify.

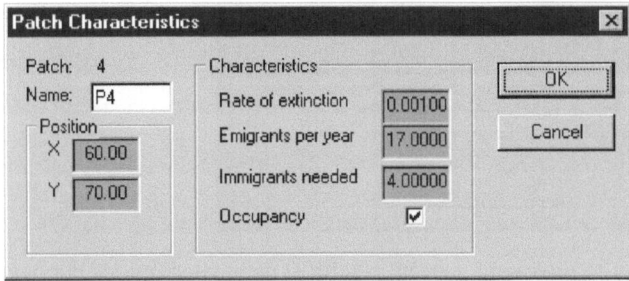

2. You can change the following parameters:

- Name of the patch.
- Position, i.e. X and Y coordinates of the patch's centre.
- Rate of extinction of a subpopulation inhabiting this patch.
- Emigrants produced per year.
- Immigrants needed to establish a new subpopulation with a probability of 50%.
- Patch occupancy (toggle the checkbox).

Changing the species-specific landscape parameters

1. To change the mean dispersal range, place your mouse cursor over the right (and bold) delimiter of the scale indicating the dispersal range.
2. Drag the delimiter to a new value.
3. To change the mean correlation length, place your mouse cursor over the right (and bold) delimiter of the scale indicating the correlation length.
4. Drag the delimiter to a new value.

9.4 Saving Your Work with the Apply Button

It is important to note that all modifications or creations performed with the Landscape Editor apply only to a copy of an existing scenario or, if you have created a new scenario with the Landscape Editor (see section below), even to a completely new scenario which is not yet saved to your disk. In order to make your modifications permanent, i.e. to save them, you have to click the **Apply** button.

Should your scenario contain user-defined matrices (indicated by the 'user icon' in the tag of the scenario in the Project Tree) you will be warned that if you continue saving your modifications, all user-defined entries in the correlation, reachability or colonization matrix will be overwritten with values calculated from the standard model:

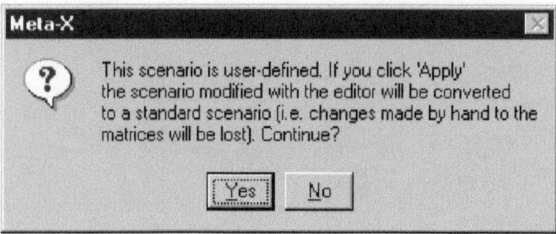

If you do not want to overwrite your user-defined parameterizations but neverthe-less want to save your work done with the Landscape Editor:

• Click **No**,
• Make a copy of the original version of the scenario (i.e. copy and paste, see Chap. 6), and then
• Click the **Apply** button. Then you should rename either the original or the new scenario and note the changes in the documentation of the scenario.

If you do not want to make your modifications introduced with the Landscape Editor permanent, simply carry on with the next work you want to perform. As soon as you select another scenario to be displayed with the Landscape Editor, you will be warned that you will lose the modifications if you do not save them:

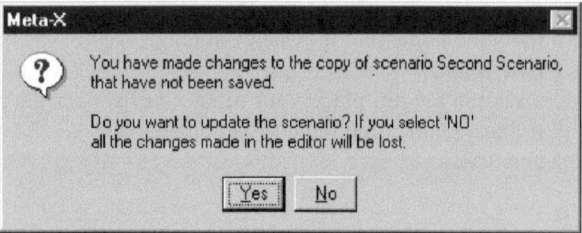

Note that again, if you chose **Yes**, all user-defined parameterizations will be lost (see above).

9.5 Creating New Scenarios

You can use the Landscape Editor to create new scenarios from scratch. For META-X beginners, it makes sense to follow the Scenario Wizard, but for experi-enced users or for exploratory work with META-X, creating new scenarios with the Landscape Editor is a useful option.

Creating a new scenario involves several steps which are listed below. The pro-cedures required for each step are described above, in the section 'Modifying ex-isting scenarios', and are not repeated here. Hence, before you create a new sce-nario with the Landscape Editor the first time, you should learn how to modify existing scenarios.

You can only create scenarios, not experiments, with the Landscape Editor. Thus, the coordinate system which defines a landscape and the spatial unit (m, km, or miles) still have to be specified with the Experiment Wizard.

The protocol for creating new scenarios

1. In the menu, select **Editor|New Scenario**.
2. In the window 'Select Experiment', select the experiment for which you want to create a new scenario.
3. Add Patches with the Context Menu.
4. Drag and drop these patches to where you want to have them.
5. Connect those patches you want to connect.
6. Double-click each patch and specify all patch characteristics.
7. Specify the dispersal range and correlation length. If you want to specify a correlation length of zero, you will first have to choose a value different from zero and then change this value to zero.
8. As an additional option, you can also modify the control parameters of this scenario with the context menu.
9. Click **Apply**.

The scenario will now have a default name. If you prefer a different name, you should now specify a new name using the Project Tree (Chap. 6).

10 Parameter Variation

Overview

With META-X, you want to assess how viability changes if you alter certain properties of the metapopulation under consideration, e.g. patch characteristics or dispersal range. Therefore, the variation of individual model parameters and the analysis of the resulting variation in metapopulation viability is a task you will frequently perform. META-X offers a special tool for generating 'variation experiments', i.e. experiments where one single parameter is varied in equidistant steps between a certain minimum and maximum value. The following parameters can be automatically varied in variation experiments:

- Local patch characteristics.
- Patch distances.
- Correlation lengths/correlations.
- Dispersal ranges/reachabilities.
- Rates of colonization.

10.1 Variation Experiments

To create a variation experiment, you first have to specify an experiment using the experiment wizard. This experiment must contain exactly one scenario. The parameters of this so-called 'base scenario' are the default parameter set of your variation experiment. Alternatively, you can take an existing experiment and delete all the scenarios in this experiment except one.

Creating Variation Experiments (1)

Parameter Variation|New opens the window 'Select Existing Experiment' which lists all the experiments in your project containing exactly one scenario. Select one of these experiments and **Continue**. The window 'Select Category of Parameter' opens:

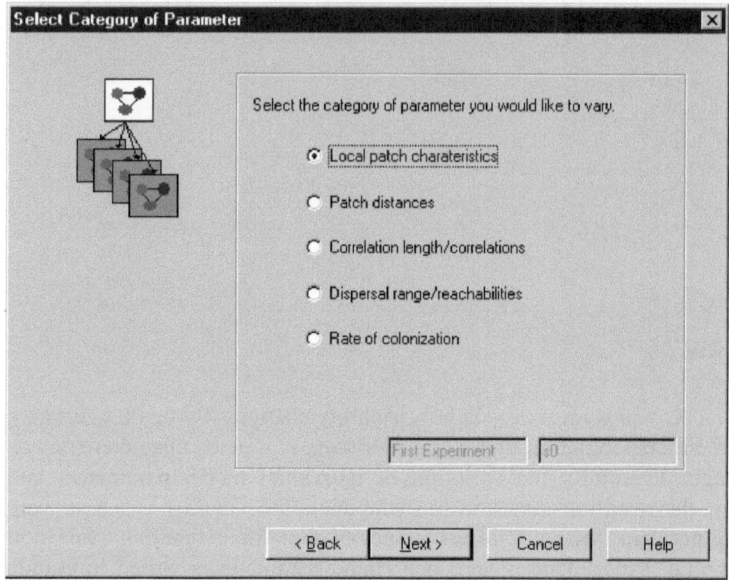

Depending on the category you choose, the following steps have to be performed (and completed by a click on **Continue**):

Varying local patch characteristics

1. Select which single patch characteristic you want to change by clicking on the corresponding cell in the matrix of patch parameters or selecting one of the three homogeneous parameters 'Rate of extinction', 'Emigrants', 'Immigrants needed' ('homogeneous' means that the chosen parameter will be the same for all patches).
2. Specify the parameter variation by specifying range of the parameter variation, i.e. start and stop value, and the number of variation steps (including the start value: to have the values 1,2,..,10 between start = 1 and stop = 10, you need 10 steps).

Varying patch distances

1. Specify the 'homogeneous patch distance' between all patches.

Varying correlation lengths/correlations

1. Select if you want to vary mean correlation length d_0 or the homogeneous correlation.
2. Specify the start and stop value of the parameter variation, and the number of steps in the variation.

Varying dispersal range/reachabilities

1. Select for which pair (i,j) of patches you want to vary the parameter 'reachability' (the probability that an emigrant from patch i reaches patch j), or select the parameter 'mean dispersal range d1' or select the 'homogeneous reachability'.
2. Specify the start and stop value of the parameter variation, and the number of steps in the variation.

Varying rates of colonization

1. Select for which pair (i,j) of patches you want to vary the main model parameter 'rate of colonization' (the probability per year that patch i successfully colonizes patch j so that a new subpopulation becomes established) or opt to vary the 'homogeneous rate of colonization'.
2. Specify the start and stop value of the parameter variation, and the number of steps in the variation.

Creating Variation Experiments (2)

The window 'Generate Scenarios for Parameter Variation' lists the choices you made for the variation experiment. If you want to modify these settings, click **Back**, otherwise click **Apply (Generate Scenarios)**. A list of names of the scenarios that have been generated will be displayed.

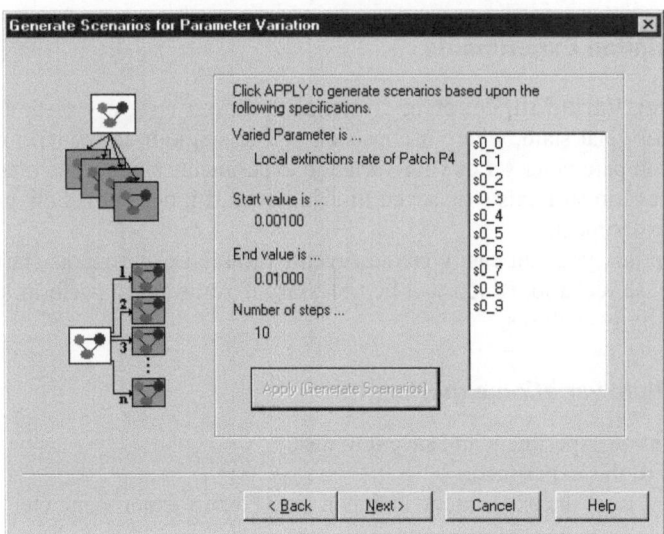

1. Click **Next**.

In the window 'End of Parameter Variation Specification', choose if you want to directly proceed to the simulation wizard to simulate the variation experiment or if you want to return to the main program.

If you choose 'Return to main program' and **Next**, click **Finish** in the next window. This will save your experiment to the Project Tree and, with **File|Save**, to your hard disk. You can now check, using the Project Tree, whether your variation experiment has been specified properly.

Deleting variation experiments

With **Parameter Variation|Delete**, you can delete variation experiments, but no other experiments. A variation experiment is an experiment which has been generated by **Parameter Variation**. This option is useful if, after some exploratory work with variation experiments, you want to delete those variation experiments no longer needed but want to make sure that no other experiments are deleted.

Modify Variation Experiments

Select this option if you want to modify an existing variation experiment. You may then modify the start and stop values of the parameter that is varied in that experiment, and the number of steps in the parameter variation.

Resetting Variation Experiments

With **Parameter Variation|Reset to 'normal'**, you can reset a variation experiment to its original state, where it consisted of one single base scenario containing the default parameter set of your variation experiment. This means that the original base scenario still exists (is saved to disk), even if it does not show up in your variation experiment.

Using this option, you can easily create several variation experiments starting with the same base scenario, which will be necessary if you want to perform local and global sensitivity analyses:

Creating multiple variation experiments

1. Create a variation experiment and save it to disk.
2. Make a copy of this experiment.
3. With the copy, reset the experiment, and then create a new experiment, etc.

10.2 Simulation, Evaluation and Reports

The options **Simulate**, **Evaluate** and **Report** of the menu **Parameter Varia-tion** work in just the same way as the corresponding options of the menu Experiments. Please refer to Chapter 11 for more help.

11 Import, Export and Report

Overview

META-X has four ways to communicate with 'the rest of the (computer) world':

- *Import of model parameters*. This is useful if you want to import model parameters from other programs, for example programs which calculate 'Patch characteristics' (mean time to extinction, etc.). The format of the import files is thus an interface between all kinds of sources which produce model parameters for all kinds of species and landscapes, and META-X.
- *Export of simulation results*. Raw data from the simulations may be exported and analyzed with other programs.
- *Export of graphs*. Some of the diagrams of META-X and the entire window of the Landscape Editor can be exported to other programs via the clipboard, or sent directly to the printer.
- *Reports*. You can generate reports of scenarios and whole experiments which contain all the information and parameters specified by you, and all evaluations. Reports are HTML format and can thus easily be imported into word-processing programs, printed and modified.

11.1 Import of Model Parameters

During the specification of a scenario, you can choose 'Read from file' to import the following parameters from file:

Parameter	Window of the Scenario Wizard
Patch positions	Position of Patches
Patch parameters (rate of extinction, emigrants per year, needed immigrants)	Patch characteristics
Correlation matrix (correlation of extinctions between pairs of patches)	Correlation length (Direct/Indirect)

Reachability matrix (probability that an emigrant from patch i reaches patch j)	Mean Dispersal range (Direct/Indirect)
Colonization matrix (rate with which patch j is colonized from patch i)	Rates of Colonization

'Read from file' prompts you to open an ASCII file (with the extension '.dat') which is in a certain format. You can import all parameters from the same file, or from different files.

Format of Import Files

Every block of parameters begins with a keyword starting with '#'. It is followed by the parameter values which are arranged in a sequence corresponding to the matrix structure of the input templates in the Scenario Wizard. The number of parameters is determined by the number of patches of the scenario:

```
% This file is for a scenario with 4 patches
#Positions
37      12
1.1     3.4
23.43   4.12
70.2    80

#LocExtinction
0.2
0.02
0.01
0.05

% column 1: Emigrants per year; column 2: Immigrants needed
#PatchCharacteristics
0.3     0.4
?       0.3
2.3     5
0.3     0.01

#Occupancy
1 0 1 1

#Reachability
0.12    ?       0.23
        0.996   0.85
                0.53

#Correlation
0.11    0.18    0.423
        0.999   0.534
                0.956

#Colonization
        0.123   0.234   0.142
0.345           0.456   0.231
1.21    0.56            0.21
9.92    1.1     0.01
```

Note that input files will often contain only a subset of the parameters shown here. In the example file, tabs have been used as separators between the parameter values, but you may also use any number of spaces. However, we recommend using tabs and mimicking the matrix structure graphically in order not to confuse the number and sequence of values to be imported. Lines starting with '%' are ignored. Missing parameters (i.e. parameter values which are not yet known) are indicated by '?' and are imported as unspecified.

11.2 Export of Simulation Results

To export raw data from simulations to ASCII files, use **File|Export Data** in the menu. The window 'Export Data to Files' will open:

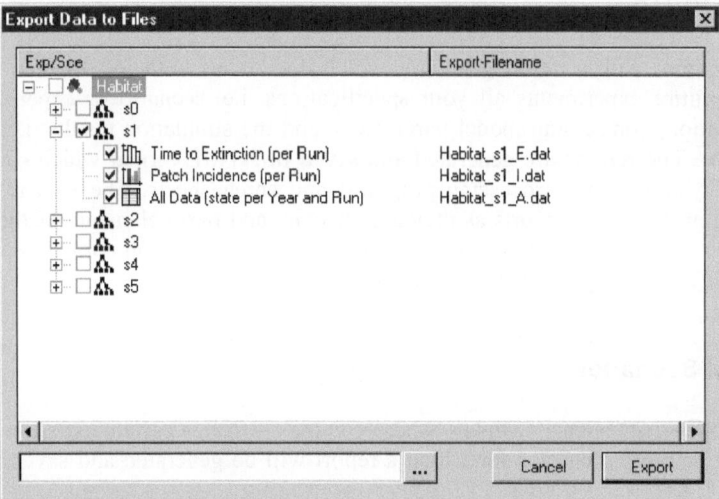

This window contains a simplified project tree whose 'branches' are expanded or packed by a mouse-click on the '+' or '-' symbols. Select by clicking the corresponding checkboxes either an entire experiment or one or more scenarios. Three different files may be generated per scenario:

- ***E.dat***: Contains in each line the number of a run and the time of extinction (or the length of the time horizon if the metapopulation did not go extinct).
- ***I.dat***: Contains in each line the number of a run and the incidence of each patch.
- ***A.dat***: This file will only contain information if for the automatic simulation of the scenarios you decided to record 'detailed information'. In this case, the file contains in each line the number of the run, the simulation year, the number of occupied patches, and the 'state', i.e. the presence/absence pattern in each year. This pattern is interpreted as a dual number, e.g. '110' means that patches 1 and 2 are occuied and 3 is empty. In the variable 'state', these dual numbers

are encoded as decimal numbers, e.g. '110'=6, '010'=2, etc. This has been done to save space in the exported file.

The filenames are formed automatically from the names of the experiment and the scenario and one of the three suffixes listed above. Specify the directory where to save the export files by clicking the '...' button.

Note: The 'All data' file may be very large if the mean time to extinction of the metapopulation is large, because then the data of very many simulation years are stored. Likewise, recording detailed data during automatic simulations may blow up your project file. Hence make sure you only record detailed data when you really need them!

11.3 Reports

META-X comes with a report generator which documents both for selected scenarios or entire experiments all your specifications, i.e. scenario or experiment documentation, control and model parameters, and the simulation results, i.e. the evaluations. The reports are generated and saved in HTML format which can be imported by most word-processing programs and spreadsheets. The idea is that you either print out the reports as they are, or copy and paste elements of the reports to assemble reports of your own (note that you directly can copy and paste most of the diagrams produced by META-X).

Report of Scenarios

Scenario|Report opens a window where you may – from a certain experiment – select one or more scenarios for which a report will be generated and saved in a file.

In the next window, you can specify a different directory and filename if you click **Change file**. The structure of the report for each scenario is:

```
Report of Scenario 's0'
1 Documentation of scenario
2 Patches of scenario
    2.1 Patch positions
    2.2 Patch distances
    2.3 Patch characteristics
2.4 Patch connections
3 Main model parameters
    3.1 Correlation matrix
    3.2 Reachability matrix
3.3 Colonization matrix
4 Control parameters
5 Evaluation
    5.1 Summary
```

Report of Experiments

The generation of reports works the same as for scenarios: **Experiment|Report** prompts you to select an experiment and then to specify a filename for the report. In the report generated, reports of all scenarios are compiled which list the same information as reports of individual scenarios. In addition, the documentation of the experiment, the environment, control parameters and the comparative evaluation of the experiment (if available) are listed. The structure of the report is thus:

```
Report of Experiment 'First Experiment'
1 Documentation of experiment
2 Environment of experiment
3 Control parameters
4 Scenarios of experiment
    4.1 Report of Scenario s0
    4.2 Report of Scenario s1
5 Evaluation
    5.1 Summary
```

If you want to have a report only of the comparative evaluation and not of the individual scenarios, click the checkbox 'Do not include scenario reports'.

Report of Projects

Choose **File|Project** to generate reports of the whole project or of selected experiments.

11.4 Exporting Graphs to the Clipboard

The following diagrams produced by META-X can be exported via the clipboard or sent directly to the printer:

- The diagrams of the scenarios evaluation, and
- The diagrams in 'TM' and in 'P0(t0)' of the comparative evaluation of experiments.

Place the mouse cursor over the diagram and click the right mouse button: choose **Graph to Clipboard** to copy the diagram to the clipboard, or **Print Graph** to print the diagram.

Likewise, you can export the map of the Landscape Editor via **Editor|Copy Bitmap to Clipboard**. You can print this map using **File|Print...** from the menu (use **Print Preview** to preview what you print).

12 Reference

Overview

In this reference chapter, we give an overview of

- The procedures of the META-X menu.
- All context menus.
- The most important warning messages in META-X.

We will not describe here in detail the procedures which are linked to the menus because they are already described in other chapters. Please follow the cross-references or use the table of contents and index to find detailed descriptions of the META-X procedures.

12.1 The META-X Menu

The menu is the backbone of META-X. It provides all the basic procedures of META-X, although many procedures can also be started via the Wizards, the Project Tree or the Landscape Editor. The main menu of META-X contains the following items:

Item	Description
File	Open, save or create files, which are called 'projects' in META-X, import and export files, and print results
Edit	Cut, copy and paste scenarios
View	Modify the view of META-X
Experiment	Working with experiments
Scenario	Working with scenarios
Parameter Variation	Working with parameter variation experiments
Editor	Working with the landscape editor
Tools	Some basic META-X options
Window	Arranging and choosing project windows

Help   Invoking the online help of META-X

Note: As with any programs running under Windows, you can use **shortcuts** to work with the menus, i.e. type the underlined character of a menu item instead of working with the mouse. If you are not familiar with shortcuts, refer to the online help of your Windows operating system.

The File Menu

The File menu contains the following elements:

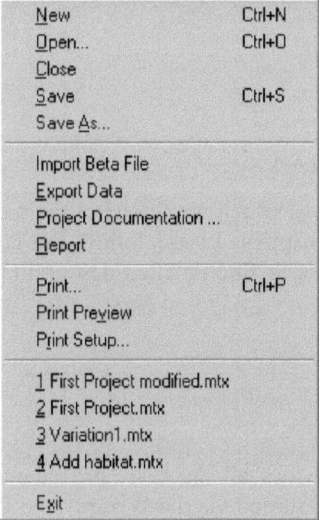

Note: File and Project are synonyms in META-X, i.e. you will save and load projects to and from disk.

There are five blocks of items in the File menu:

File management of projects

Item	Description
New	Create a new project.
Open	Load an existing project from disk.
Close	Close the currently open project. Make sure that you have already saved this project in order to save your work with the project!
Save	Saves the project to disk. We strongly recommend – as we would

for any program on computers – saving your work regularly and not working for hours without saving your work.

Save As   Saves the project to disk. You may specify a new file name. This option is useful if you want to create copies of existing projects.

Export (and import)

Item	Description
Import Beta File	This option is only for those who tested the beta release of META-X, which used a different file format.
Export Data	Exports raw data from the simulations.
Project Documentation	Opens the Project Documentation for inspection and editing.
Report	Starts the generation of reports of experiments.

Printing

Here, **Print...** prints the window of the Landscape Editor (copy the window to the clipboard before using **Editor|Copy Bitmap to Clipboard**), **Print Preview** allows what you are going to print to be previewed, and **Print Setup** is the standard Windows procedure for printer setup. For other ways to export items from META-X, see Chapter 11.

History

The fourth block of the File menu is a list of the four most recently opened projects. Clicking on one of these projects or typing the corresponding number will load the project.

Finally, with **Exit** you can finish the current session, exit META-X and return to the operating system.

The Edit Menu

The edit menu comprises four items:

Item	Description
Undo	Not implemented.
Cut	Cuts the scenario which is selected in the project tree and copies it to the clipboard.
Copy	Copies a scenario which is selected in the project tree to the clipboard.
Paste	Pastes a scenario from the clipboard into the experiment whose

'Scenario' item is selected in the project tree.

The View Menu

Here you may choose to view (set by default) the toolbar and the status bar. The toolbar is the bar of buttons directly underneath the main menu. These buttons allow direct access to frequently used procedures, without having to use the menu. If you place the mouse cursor over a button, its procedure is described in a hint that pops up.

The Experiment Menu

This menu has the following six items:

Item	Description
New	Start the Experiment Wizard and create a new experiment.
Modify	Select an existing experiment and start the Experiment Wizard.
Delete	Select an experiment and delete it.
Simulate	Simulate an experiment.
Evaluate	Display the evaluation of an experiment.
Report	Generate a report of an experiment.

If you are working with the Landscape Editor, the Experiment menu will not be active. Click somewhere in the window of the Project Tree to activate the menu.

The Scenario Menu

This menu has the following nine items:

Item	Description
New	Start the Scenario Wizard and create a new scenario.
Modify	Select an existing scenario and start the Scenario Wizard.
Delete	Select a scenario and delete it.
Interactive Smulation	Start interactive simulation for a selected scenario.
Automatic Simulation	Start the automatic simulation for a selected scenario.
Evaluation	Display the evaluation of a selected scenario.

Report	Generate a report of a selected scenario.
Move up	Move a selected scenario up in the list of scenarios in the Project Tree. A scenario must be selected in the Project Tree to use this option
Move down	Move a selected scenario down.

If you are working with the Landscape Editor, the Scenario menu will not be active. Click somewhere in the window of the Project Tree to activate the menu.

The Parameter Variation Menu

The Parameter Variation menu is for a special type of experiment: variation experiments. A variation experiment is an experiment which consists of a base scenario and a series of scenarios which are generated from the base scenario by just varying one single parameter. Therefore, this menu has the same items as the Experiment menu, except 'Reset to 'normal'', which means resetting a variation experiment to its original state where it consisted of only one base scenario.

The items of the menu will only be active if your project contains variation experiments. Otherwise, only the item 'New' will be active.

The Editor Menu

This is the menu of the Landscape Editor. It has 14 items:

Item	Description
New Scenario	Create a new scenario with the Landscape Editor.
Load Scenario	Load a scenario to the Landscape Editor.
Apply	Make your additions or modifications of the scenario permanent in the project (but not yet on disk; for this, choose **File|Save**).
Local Aspect	Choose the patch characteristic, i.e. mean time to extinction, which is to be used to scale patch size in the map of the Landscape Editor.
Add Patch	Add a patch to the landscape.
Delete Patch	Delete a patch (a patch must be selected by a mouse-click).
Connect	Connect a pair of patches.
Disconnect	Disconnect a pair of patches.
Set occupied	Set a selected patch occupied (by a subpopulation).
Set empty	Set a selected patch empty.
Toggle Grid	Toggle a grid underlying the map.
Copy Bitmap to Clip-	Copy the window of the Landscape Editor to the clipboard

| board | (and paste it into another program). |

The Tools Menu

This menu has two sub-menus: **Options** and **Customize**. The **Options** menu contains two tagged panels.

Default Options for View

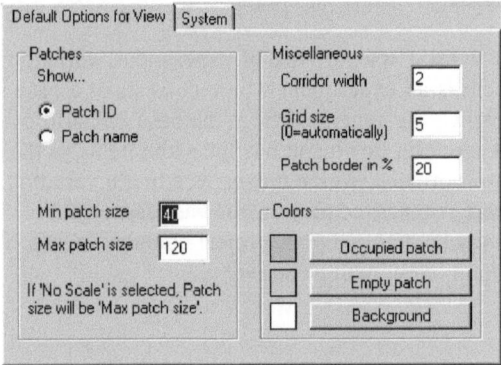

The options are:

Option	Description
Patch ID/Patch name	Determines whether the patch identity number or its name are displayed in the Landscape Editor.
Min/Max patch size	Determines the range of sizes (in relative units) of the patch symbols. The range is confined to the interval [35, 160].
Corridor width	Width of the lines representing connections between patches. Values must be within the interval [0, 7].
Grid size	Spacing of the grid lines (see **Toggle Grid**).
Patch border	Width of the border of the patch symbols (whose colour indicates whether the patch has been completely parameterized; green: complete; red: incomplete).
Colours	Here you can choose colours for the symbols of occupied and empty patches and for the background of the landscape map.

System

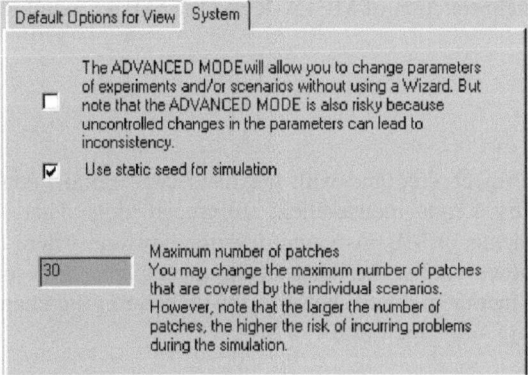

The options are:

1. Checkbox 'Advanced Mode'. If this mode is activated (marked in the checkbox), you can edit the parameter overviews which you invoke from the Project Tree. You thus do not have to run through the Experiment and Scenario Wizard any longer, but can directly choose and edit parameters. However, this may lead to inconsistencies with user-defined scenarios, where elements of the parameter matrix are not calculated by the META-X submodels but are specified by you. If you use the advanced mode, make sure you check your scenarios from time to time by inspecting them with the Scenario Wizard.
2. Checkbox 'Use static seed for simulation'. META-X uses computer-generated random numbers to mimic random fluctuations. If you use 'static seed', the sequence of random numbers is the same for each single simulation because it is started with the same number ('seed').

Customize

Enables the tool bar to be customized.

The Window Menu

This menu is mainly the standard menu of the Windows operating system to create copies of the existing window (**New**) or to arrange the windows of several projects. The item 'Project Workspace' indicates whether you are currently working with the Landscape Editor (Project Workspace item active) or not (inactive). In the former case, the Experiment and Scenario menus are not active. Click in the area of the Project Tree to deactivate Project Workspace. Finally, the Windows menu lists the currently open projects. To switch to one of these projects, click on the project's name in the list, or type the corresponding number.

The Help Menu

The Help menu invokes the help function of META-X and the 'About' window.

12.3 Context Menus

For working with both the Project Tree and with the Landscape Editor, context menus, which are activated by a right mouse-click, are crucial tools. Therefore, here we list all the context menus of META-X and describe how and where they are activated. We will not, however, list the items of the menus here or describe the procedures linked to the menus in detail, because this is done in the chapters about the Project Tree and the Landscape Editor.

Project Tree

In the following, the items (or branches) of the Project Tree and their context menus are listed.

'Experiments':

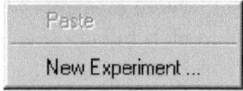

Name of an experiment:

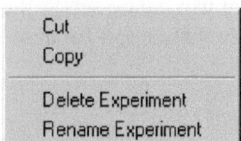

'Experiment Documentation' (or the next two items in the experiments branch):

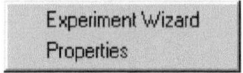

'Scenarios':

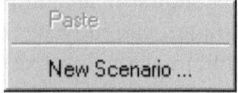

A scenario's name:

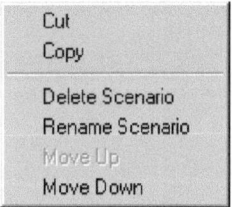

'Scenario Documentation' or one of the other items of the scenario's branch (except 'Evaluation (of Scenario)'):

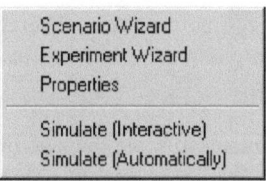

Note: The option 'Simulate (Interactive)' and 'Simulate (Automatically)' will show up only if the scenario is completely specified. Incomplete scenarios cannot be simulated.

Landscape Editor

The Landscape Editor has one, basic context menu which is activated by a right mouse-click anywhere within the workspace of the Landscape Editor:

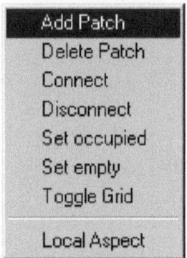

Note: Some of the options are only available if you have selected a patch.

The Landscape Editor also has two menus which are invoked by a double-click with the mouse.

If you double-click a patch, you can inspect or edit the parameters of this patch:

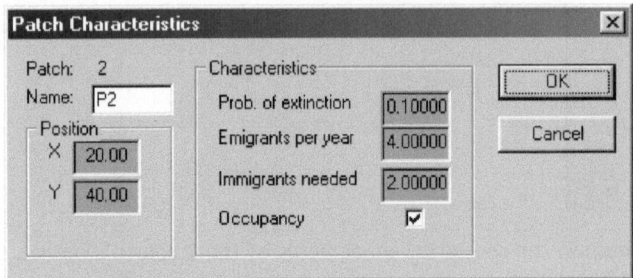

Note: If you choose **OK**, the scenario will be marked as modified. Therefore, choose **Cancel** if you want only to view the patch parameters but not to edit them.

If you have selected a Local Aspect, i.e. a patch parameter to scale the size of the patch symbols (usually this will be the patch's mean time to extinction), double-clicking on the patch symbol underneath the button **Apply** opens the Patch Scaling menu:

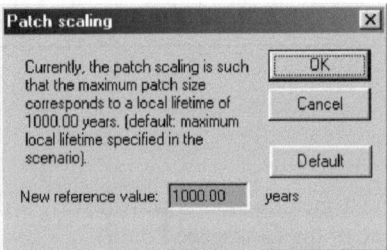

Evaluation

The evaluation diagrams of a scenario or of the comparative evaluations of experiments (tags **TM** and **P0(t0))**, are linked to the context menu:

which allow the diagram to be printed or copied to the clipboard.

12.4 Warning Messages

There are several warning messages which now and then will pop up while you are working with META-X (Don't panic!). The most important warning messages are listed and explained below; the warnings are also explained in other parts of the manual.

Scenario Wizard Warning Messages

The 'reset' warning:

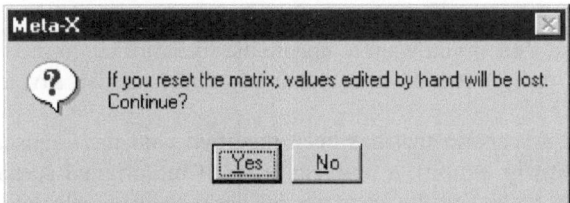

This message appears when you push the button **Reset Matrix** in those windows of the Scenario Wizard where the reachability and correlation matrix are displayed. If you have edited matrix elements by hand, or read matrix elements from file, a 'reset' will replace all these 'user-defined' values by values which are calculated by the META-X standard submodels. Thus, make sure you do not 'reset' matrices unintentionally.

The 'colonization matrix' warning:

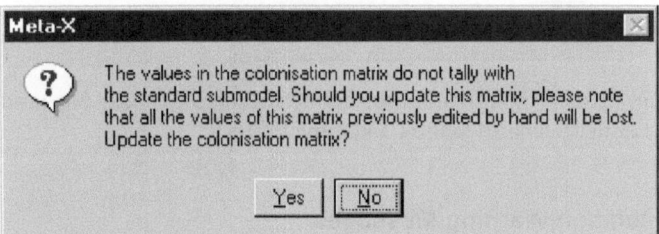

Make sure that you choose **NO** if you are working with user-defined scenarios.

Landscape Editor Warning Messages

If you have modified a scenario with the Landscape Editor and try to select another scenario in the Project Tree of the menu without having clicked the **Apply** button, you will be warned that all your changes might be lost:

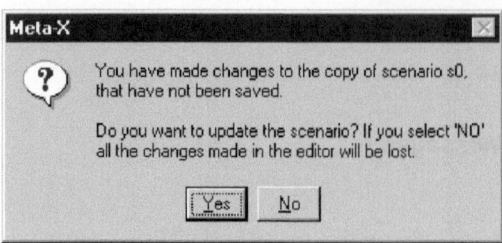

Make sure that you choose **Yes** if you want to update the scenario, i.e. to save the modified scenario to the project (which does not imply that it is also saved on disk!).

If you want to update a scenario that has been modified with the Landscape Editor (by clicking the **Apply** button), a problem arises if the original scenario was user-defined. In order to update the scenario, META-X has to recalculate all the main model parameters, and this can only be done automatically using the standard submodels of META-X.

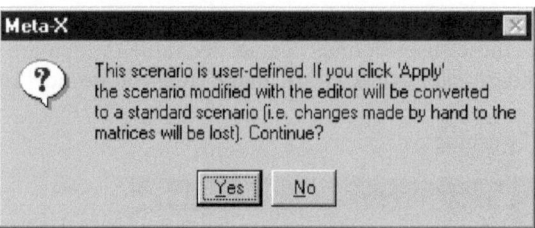

Hence if you want to keep the original user-defined scenario, choose **No** (and instead load a copy of that scenario to the Landscape Editor)

Interactive Simulation Warning Messages

During Interactive Simulation, you may modify the scenario that you loaded for simulation (using the Wizard or – if you were in Advance Mode – the Properties button). When you return to the main program, you will be asked if you want to save these changes:

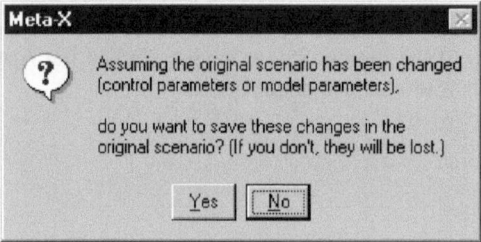

Be careful to choose **Yes** if you want to save the changes, otherwise select **No**.

13 Goals, Methods, and Concepts of PVA

13.1 Introduction

META-X has been developed as a tool for education and for practical use by specialists such as field ecologists, conservation biologists, managers of natural resources, and environmental decision-makers. Working with META-X requires no programming and – at least at the level of the metapopulation – no modelling, and is therefore easy to use by non-experts. However, as with any tool, even non-experts have to know the purpose of the tool and its basic function if they want to use it appropriately. Moreover, META-X is not a fully 'canned' PVA software because external sub-models, for example of local population dynamics, are needed to parameterize META-X for actual species and landscapes. Therefore, to work with META-X you need to be familiar with the basics of PVA. In this chapter we briefly introduce the goals, methods and concepts of population viability analysis (PVA). In particular, we explain in detail how to quantify the persistence and viability of populations (and metapopulations), since this is usually not explained in PVA literature. For those needing a full introduction into PVA, we strongly recommend consulting the literature listed at the end of this chapter.

In the following, we first introduce the goals of PVA and the tool that makes PVA quantitative: ecological models. Then, using a very simple stochastic population model, methods of quantifying persistence and viability are demonstrated in detail. Afterwards, PVAs of real species and the general potentials and limitations of PVA are briefly discussed. This is important because many different views exist on the role of PVA, ranging from over-optimistic to over-pessimistic (for an overview, see Burgman and Possingham 2000). Finally, we introduce specific elements of the PVA of metapopulations and then discuss alternatives to tailored PVA models and explain the conception of META-X.

13.2 Goals of PVA

The basic question of PVA is: 'What are the minimum conditions for the long-term persistence and adaptation of a species or population in a given place?' (Soulé 1987, p.1). A viable population is a population which has the capacity to "maintain itself ... for the foreseeable ecological future (usually centuries) with a

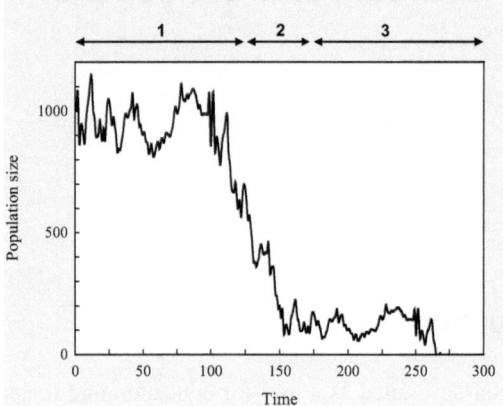

Fig. 13.1. Three typical stages in the development of a population that has gone extinct. The third stage is the main realm of PVA.

certain, agreed on, degree of certitude, say 95%" (Soulé 1987, p.2). PVA has been regarded as a subdiscipline of conservation biology since the mid-1980s. Key publications include Shaffer (1981), who introduced the concept of 'minimum viable population' (MVP) and the 'blue book' entitled *Viable populations for conservation*, which was edited by Soulé in 1987 (see Quammen, 1996, for a historical and personal account of PVA).

The establishment of PVA reflects the growing concern over the dramatic loss of species due to anthropogenic impact and over the even more dramatic prospect of future species loss on the scale of the mass extinctions recorded by fossils. It has been predicted that "if current land-use practices continue, about half of all species now living will be lost within the next 50 to 100 years" (Burgman et al. 1993, after Myers 1981, Simberloff 1986, May, 1988, 1990, Wilson 1988).

To understand the goals of PVA, it is important to distinguish between three stages in the fate of a population that has gone extinct (Fig. 13.1). In the first stage, the population is viable, i.e. so large that there is – as far as can be foreseen – no concern over extinction. The population size may fluctuate or cycle, but these fluctuations are considered largely irrelevant for persistence. In the second stage, the population is on average declining. The agents responsible for this decline may be natural, but nowadays are most commonly anthropogenic, e.g. habitat loss and fragmentation, over-exploitation, pollution, or the invasion of alien species. If the trend of the decline were to continue, the population would go extinct more or less deterministically. However, in many cases the agents responsible for decline will cease before the population is extinct. For example, remnants of habitat which are unsuitable for human use (e.g. steep mountain slopes) may remain; exploitation may become non-profitable at low population densities, etc.

Thus, many declining populations enter the third stage (Fig. 13.1), which is similar to the first stage in that it shows no average decline. However, now the population is so small that random fluctuations of the environment and of demographic processes within the population cause fluctuations of population size which may reach the order of magnitude of the population size itself and therefore, sooner or later, may lead to extinction. This third stage – but also the second stage if the population growth rate remains negative – is the realm of PVA. Thus, the proximate reasons of extinction addressed by PVA are random fluctuations, whereas the ultimate reasons are, of course, the agents which lead to decline (Simberloff 1986, Caughley 1994). The question then is: Is the population too small, i.e. is the risk of extinction over a certain time horizon of, for example, 100 years greater than acceptable, e.g. larger than 5%?

The proximate goal of PVA is the quantitative assessment of the risk of extinction (Burgman et al. 1993). The ultimate goal, however, is to enable management for viable populations. Typical questions include: How large must a habitat be to allow for viable populations? How effective are different management alternatives which try to re-establish viability? What natural mechanisms allow for viability and how can these mechanisms be mimicked by management?

PVA is based on *quantitative* risk assessment because verbal classifications of populations as being 'too small' or 'highly endangered' are too vague to be useful in environmental decision-making. In PVA, risk assessment is quantitative because ecological models are used which are implemented mathematically or, more frequently, as computer programs. Thus, performing PVA requires understanding the rationale of ecological modelling.

13.3 Ecological Models for PVA

Models are – more or less consciously – used all the time when a problem has to be solved or a question answered. Models are purposeful representations (Starfield et al. 1990) which allow efficient solutions to problems to be developed and tested. If we had no models, we would have to stick to trial-and-error all the time. Representation may be verbal, graphical, or use the language of mathematics and computers, but the essential aspect of models is independent of these media: a model should contain only those aspects of the real system that are considered essential to the problem. The reason for this is the principle of parsimony: the simpler a model, the easier it is to develop, test, understand and communicate. On the other hand, models are necessarily simpler than their real counterparts because knowledge about the real system is necessarily limited. Problem-solving and, in turn, modelling are always subject to constraints of information and time. Modelling means compromising for these constraints and making the best of the information and time available. This is in particular true for models used in PVA, because conservation biology is a 'crisis discipline' which definitely cannot wait until sufficient data are available.

Technically, modelling for PVA requires choosing state variables (e.g. population size, or number of individuals with certain attributes, such as age, sex, size, fecundity, social rank), temporal resolution (years, weeks, days), spatial resolution (none, or some resolution pertaining to the species and processes involved), process resolution (birth and death rates or using submodels of behaviour, life cycle etc. which produce birth and death rates as a result), structural resolution (e.g. homogeneous or heterogeneous habitat), and which elements of the model should be deterministic or stochastic. Each of these choices depends on the information available and the element's significance. For example, age structure is certainly important, but if no data on age-dependent survival and birth rates exist, there is no point in building an age-structured model. On the other hand, detailed knowledge about, for example, lekking behaviour should be ignored in the model unless it is considered essential for population dynamics and, in turn, extinction risk. Likewise, descriptions of dispersal have to be extremely aggregated by using dispersal rates, unless detailed knowledge about dispersal behaviour and landscape structure is available. And finally, although random fluctuations will affect virtually all aspects of population dynamics, much of this variation may be ignored as mere noise. However, fluctuations which cause strong variations in the population's growth rate must not be ignored in models addressing the viability of populations.

In PVA, both deterministic and stochastic models are used. Deterministic models in PVA are usually matrix models which describe age- or stage-structured populations (Caswell 1989, Burgman et al. 1993). The purpose of these models is to determine the deterministic intrinsic rate of increase (growth rate), r_m, of populations and how r_m depends on demographic parameters. If it turns out that a population has a negative growth rate, deterministic models may help develop management strategies to make r_m positive. If, however, r_m is zero or positive, deterministic models are no longer sufficient (Burgman et al. 1993) and stochastic models are needed, i.e. models that take random variations into account. Deterministic models are most useful in the second stage described in Fig. 13.1, whereas the realm of stochastic models is the third stage.

13.4 A Stochastic Example Model

A wide variety of stochastic population models is used in PVA, ranging from very simple, highly aggregated models to extremely detailed ones. Nevertheless, the backbone of many of these models (including – at the metapopulation level – the META-X model) is very similar in form and can be demonstrated with a model developed by Burgman et al. (1993). Originally, this model was designed for a small population of white rhinos in a small reserve, but we will simplify the model even more and use arbitrary parameters. Thus, the following example is about a hypothetical species. First, the three most important elements of this and all other stochastic population model for PVA are described in detail.

1st Element: Demographic Noise

The basic idea is to use computer-generated random numbers to simulate *demographic* and *environmental noise*. Demographic noise is the variation of the population's growth rate which is due to the fact that the fate of individuals is – to some degree – independent. One individual may do well and reproduce, another may be eaten by a predator, and still another may fail to reproduce because it has not found a mate. All these random variations are ignored when we consider the average birth or death rates of a population. Birth and death rates are determined empirically by averaging the birth and death processes observed of all individuals. For example, if in a certain year 100 individuals produce 20 offspring, the birth rate is $b=0.2$. In simulation models, demographic noise is mimicked by interpreting b as the *probability* of an individual producing, for example, one offspring. For fairly large numbers of individuals, this probabilistic interpretation of b makes almost no difference to a deterministic model where the individual number would simply be multiplied by b. However, if only a small number of, say, fewer than 20 individuals is left, significant deviations from the deterministic mean may occur. This is well-known from throwing a die: 100 throws are very likely to lead to a fraction close to 1/6 for each of the six numbers possible, whereas with only 10 throws large deviations from 1/6 may occur.

In ecological simulation models, 'probability' is very easily implemented by using random numbers, z, generated by the computer. These numbers have a uniform distribution, i.e. each value between 0 and 1 has the same probability (equal to one) of occurring. To simulate birth events occurring with a probability of $b=0.2$, a random number z is generated and compared to b. If $z \le b$, a birth event occurs and the number of offspring is increased by one; if $z>b$, nothing happens.

2nd Element: Environmental Noise

However, the fate of individuals is only partly independent of each other. In good years, all individuals will have a more or less lower death rate and an increased birth rate, and in bad years *vice versa*. Environmental noise, i.e. variation in the population growth rate which is caused by random variations of the environment, partly synchronizes the fate of individuals. Typical elements of the abiotic or biotic environment which fluctuate include the weather, resource availability, the presence of predators and competitors, and epidemic diseases. In simulation models, environmental noise means that birth or death rate, or both, are no longer constant but vary from year to year.

In ideal cases, the variation of birth and death rates is known empirically (e.g. Dorndorf 1999), but such detailed information is rarely available in PVA. Often we have to rely on best guesses by experts, i.e. experts who know the population in the study area well enough to be able to assess the frequency of good and bad years (or of other classes of environmental quality if necessary) and assign values of demographic parameters to these classes (e.g. Stephan et al. 1995). Alternatively, the empirical record of an environmental factor which is known to strongly affect, for example, juvenile survival, can be used to determine a probability distribution of juvenile survival, e.g. rainfall in the period where juveniles are susceptible to cold and wet weather (as is the case with capercaillie *Tetrao urogallus*; Grimm and Storch 2000). A new, promising way of fine-tuning demographic parameters is to fit the model output of structurally realistic models (obtained using the pattern-oriented modelling approach; Grimm et al. 1996) to observed patterns in time series or of presence-absence data (Wiegand et al. 1998, Hanski 1999, Wiegand et al. 2002; Chap. 15).

3rd Density Dependence

Density dependence means that the growth rate of a population has to be zero at a certain population size or density, otherwise the population will grow infinitely. The simplest way to model density dependence is to introduce a ceiling population size K which may be envisaged as the carrying capacity of the area in question (e.g. Lacy et al. 1995, Grimm and Storch 2000). If the population size N is below K, density-dependent effects on reproduction, mortality and emigration are ignored; otherwise, if $N>K$, randomly selected individuals are removed from the model population until $N=K$. For unstructured models, where N is the state variable, differences between this ceiling approach and more mechanistic approaches may be negligible, but for structured models, for instance age-structured models, the way density dependence has been modelled largely influences the absolute assessments of extinction risk, i.e. the numerical value for a given set of parameters (Chapman et al. 2000).

Implementing the Example Model

The program, which implements the example model, consists of three nested loops. In programs, a loop is a block of actions that is performed repeatedly. The innermost loop is the N or individual's loop: for each individual, two random numbers are generated to determine whether the individual reproduces or dies (demographic noise). In the *years* loop, in each year first the birth rate is drawn from a uniform distribution, i.e. from the interval $[0, 2b]$ (which is not very realistic but sufficient to demonstrate the general idea of environmental noise; we also ignore here variations in the death rate), and then the individuals loop running over the actual number of individuals, $N(t)$, is started. The years loop is repeated until one run is finished, i.e. until the population is extinct or until a predefined time horizon of, for example, 100 years is reached. The result of the years loop is the population dynamics $N(t)$ and an extinction (or survival) time T, i.e. the time when the population goes extinct in this particular run. Due to the random processes in the model, which represent demographic and environmental fluctuations, $N(t)$ and T will vary from run to run. Therefore, for a statistical assessment of the viability of the population, many runs have to be performed in the *runs* loop. In

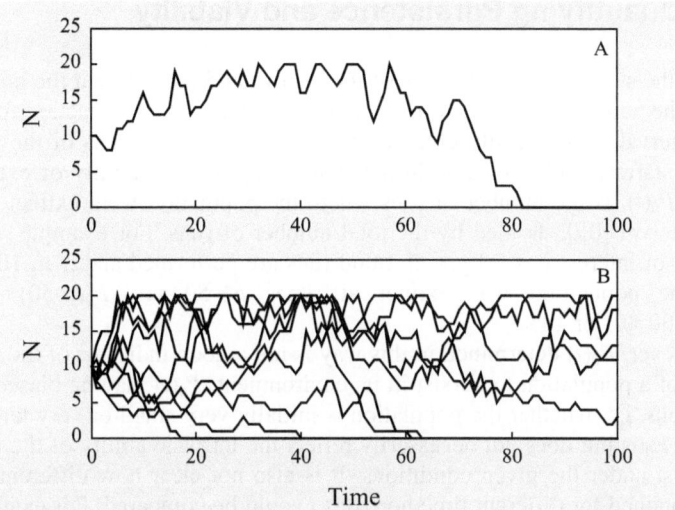

Fig. 13.2. Population dynamics produced by the example model. A: A single run. B: Nine runs which started from the same initial state N_0=10 but used different random numbers. Four populations go extinct within the time horizon of 100 years; five populations survive.

each run, first the initial state of the population is set (here, $N(t)$=N_0) and then the years loop is started.

Results of the Example Model

The results of our example model, which are typical for any stochastic population model, are shown in Fig. 13.2. The birth and death rates are chosen such that on average the population has a positive growth rate of r_m=b-d=0.05. Nevertheless, in many years the population size N goes down due to demographic and environmental noise. The population dynamics shown in Fig. 13.2a end with extinction at time T= 82 yrs. However, this particular run is not representative of the ability of the population to persist. Starting with the same initial number of individuals, different sequences of random numbers lead to other population dynamics and other extinction times (Fig. 13.2b). This variation is real. It reflects the fact that neither we nor the individuals of a real population can predict how weather or predation pressure will change, or what the fate of each individual will be.

This inherently stochastic nature of the dynamics of small populations renders deterministic predictions of population viability impossible. We can only quantify the *probability* of extinction over certain time horizons.

13.5 Quantifying Persistence and Viability

Due to the stochastic processes described above and the fact that the population is small, the population is at risk of going extinct. The risk of extinction for a certain time interval is very easily obtained from repeated simulations of the population model, starting with the same initial state. Then the probability of extinction by time t, $P_0(t)$, is the number of runs where the population went extinct within the time interval $[0,t]$, divided by the total number of runs. For example, if the time horizon of interest is $t=50$ yrs, if 1,000 runs are performed and if in 100 of these runs the population was extinct at times $t\leq50$ yrs, $P_0(t=50)$ would be 100/1,000=0.1 or 10%.

However, $P_0(t)$ determined in this way is not a good indicator of the theoretical ability of a population to persist in its environment. $P_0(t)$ may be biased by initial conditions, i.e. whether the population is initially very small or very large will influence $P_0(t)$ but does not necessarily reflect the intrinsic ability of the population to persist under the given conditions. It is also not clear how different values of $P_0(t)$ obtained for different time horizons t could be compared. For example, Marshall and Edwards-Jones (1998) determine the minimum capacity of habitat for capercaillie (*Tetrao urogallus*) which would lead to a $P_0(t)$ smaller than 0.05 or 5% in 50 years, whereas Grimm and Storch (2000) use 1% and 100 years as criteria for viability. How are these assessments of viability interrelated?

A more basic approach to quantify persistence and viability is based on theoretical insights gained from abstract, idealized models (Wissel and Stöcker 1988, Stephan 1993, Wissel et al. 1994, Stelter et al. 1997, Grimm and Wissel 2002). It can be shown that for stochastic population dynamics which start at time $t=0$, after a transient time (which is usually very short), the following relationship holds for the probability of extinction by time t:

$$P_0(t) = 1 - c_1 e^{-t/T_m} \qquad (1)$$

The two parameters of this equation, c_1 and T_m, have a clear and important ecological interpretation which will be explained below.

Established Phase and Intrinsic Mean Time to Extinction

The interpretation of the two parameters of Eq. 1 follows from the theory of stochastic processes: for most situations which are relevant in PVA, c_1 is a good approximation of the probability of the population reaching the "established phase" (see Grimm and Wissel 2002, for more detailed explanations). The established phase is characterized by typical fluctuations of the models' state variables (in our example model population size N, Fig. 13.2a). If simulations are started with state variables which are below the range of typical fluctuations, c_1 approximately gives the probability of the population reaching the range of typical fluctuation, i.e. the established phase. If, for example, the population is extremely small at $t=0$, in many simulation runs the population will obviously go extinct before it reaches the established phase. If, on the other hand, the population size is initially within the range of typical fluctuations, the population is already established and, in turn, $c_1=1$.

In the established phase, the risk of extinction per short time interval is constant (and equal to $1/T_m$; Wissel et al. 1994, Grimm and Wissel 2002). Therefore, the notion of an established phase of a population helps to distinguish between the intrinsic ability of a population to persist in a certain environment, and the effect of initial conditions, i.e. of the state of the population at the beginning of the simulation. Whether the initial state of the population belongs to the established phase can easily be checked with the $\ln(1-P_0)$ plot (see next page): it is the case if the parameter c_1 in Eq. 1 is equal to one and thus the intercept of the $\ln(1-P_0)$ plot, $-\ln(c_1)$, is equal to zero.

The other parameter of Eq. 1, T_m, is independent of the simulation's initial conditions and therefore may be referred to as "intrinsic mean time to extinction" (Grimm and Wissel 2002). Technically, T_m is determined with the $\ln(1-P_0)$ plot (see below). The two ecological interpretations of T_m are equivalent and refer to populations which are in the established phase: $1/T_m$ is the constant risk of extinction per short time interval and, in turn (as can be shown analytically), T_m is the arithmetic mean of the distribution of extinction times.

Persistence and Viability

The relationship in Eq. 1 and the concept of the intrinsic mean time to extinction, T_m, allows us to distinguish conceptually between persistence and viability, and to generalize the concept of minimum viability. T_m characterizes the intrinsic ability to persist, i.e. the *persistence*, of a population in a given environment. T_m may be rather large, e.g. 10,000 years, but this does not mean that we should use PVA models to project 10,000 years into the future (which would, of course, be absurd; Beissinger and Westphal 1998). Instead, T_m is the basic currency (Grimm and Wissel 2002) to quantify persistence, because once T_m has been determined, the probability of extinction by time t, $P_0(t)$, can be calculated for all time horizons t of interest by using Eq. 1.

However, in conservation we are focusing on situations with $P_0(t) \ll 1$. In this cases, and if $c_1=1$, the exponential function in Eq. 1, $\exp(-t/T_m)$, can be linearly approximated by $1-t/T_m$, which leads to:

$$P_0(t) \approx \frac{t}{T_m} \tag{2}$$

Eq. 2 uses the currency T_m to calculate *viability*, i.e. the ability to persist over a certain time horizon for any time horizon of interest (for larger $P_0(t)$ or for initial states which are not yet established, the full Eq. 1 has to be used instead). Persistence thus refers to the intrinsic property of a stochastic process (population dynamics), whereas viability combines this intrinsic property with our time horizon of interest.

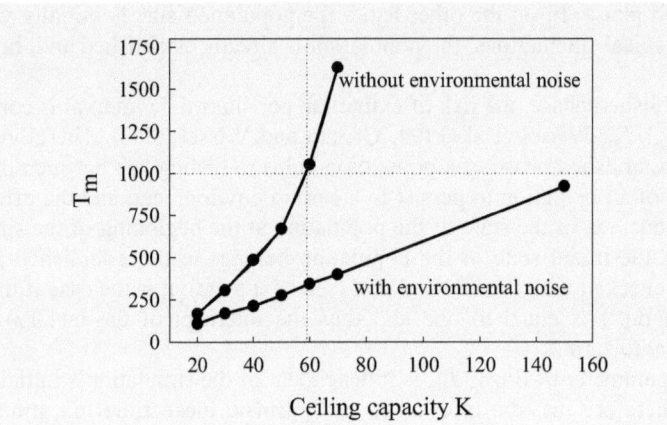

Fig. 13.3. The intrinsic mean time to extinction, T_m (years), versus the model parameter K (individuals). T_m=1,000 years corresponds to a minimum viability of 5% in 50 years. The intercept of this criterion with the plots delivers the minimum capacity required, for example approximately 60 individuals in the case where environmental noise is ignored.

Eq. 2 (or Eq. 1) can be used to compare different definitions of *minimum viability*. If our demand is that we tolerate at maximum a risk of extinction in 50 years of, for example, 5% (or 0.05), this would mean that T_m has – according to Eq. 2 – to be at least $t/P_0(t)$=50/0.05 =1,000 years, whereas 1% in 100 years would require T_m to be larger or equal to 100/0.01=10,000 years. Now, if we determine T_m in our model for different values of a critical model parameter, for example, as in Fig. 13.3, the ceiling carrying capacity of the habitat, the demand that $T_m \leq 1,000$ years allows us to determine the threshold of the critical parameter. This is easily done in a plot of T_m versus the parameter (Fig. 13.3).

The ln(1-P₀) plot

T_m and c_1 are determined by a plot which utilizes the mathematical structure of Eq. 1. We make Eq. 1 linear by taking the natural logarithm on both sides, which after some basic rearrangement yields:

$$-\ln(1-P_0(t)) = -\ln(c_1) + {t}/{T_m}$$

This is a linear relationship of the type $y=y_0+mx$. Thus, a plot of $-\ln(1-P_0(t))$ versus time t should yield a straight line with slope $1/T_m$ and y-intercept $-\ln(c_1)$. To produce this plot, the following has to be done (Stelter et al. 1997, Grimm and Wissel 2002):

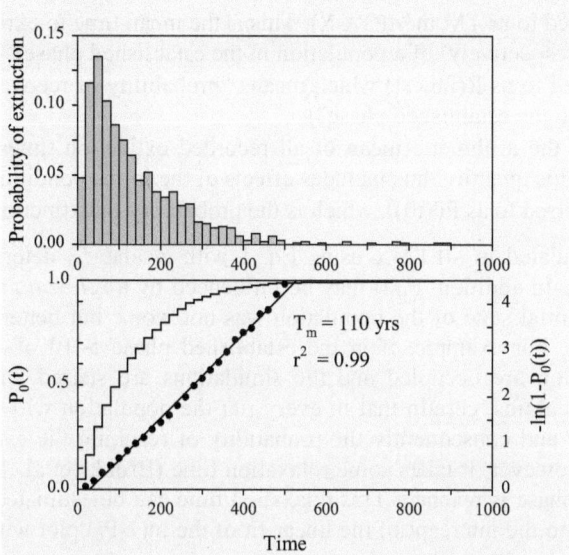

Fig. 13.4. Producing the -ln(1-P₀) plot. A: Normalized histogram. B: Probability of extinction by time t, $P_0(t)$, obtained from the histogram in A, and the plot of $-\ln(1-P_0(t))$ versus t whose linear fit has a slope of 1/110 years^{-1}.

1. Produce a histogram of the extinction times T obtained from, for example, 1,000 simulations. Normalize this histogram (Fig. 13.4), i.e. divide each bar of the histogram by the total number of runs. Then, each bar gives an estimate of the probability $p(t')$ of the population going extinct in the time interval t' which corresponds to the bar.
2. Now, the probability of the population being extinct after the time corresponding to the first bar is simply equal to this first bar; the probability of the population being extinct after the first two bars is simply the sum of the first and second bars, and so on. Thus, by cumulatively summing up the bars of the histogram, the function $P_0(t)$ is produced:

$$P_0(t) = \sum_{t'=0}^{t} p(t')$$

3. Finally, plot $-\ln(1-P_0(t))$ versus time t and perform a linear fit.
4. Determine the slope $(=1/T_m)$ and the intercept of the fitted line $(= -\ln(c_1))$.

Fig. 13.4 shows the normalized histogram and the ln(1-P₀) plot for results obtained from the example model. Technical aspects of the plot are explained in the Appendix of this chapter.

In META-X, the quantification of persistence and viability is based on the ln(1-P₀) plot. The quantities calculated in META-X are:

- T_m (referred to as TM in META-X). This is the mean time to extinction (or mean lifetime, respectively) of a population in the established phase.
- c_1 (referred to as R(ini,est) which means "probability of recovery from the initial state to the established phase").
- T_{mean}, i.e. the arithmetic mean of all recorded extinction times (referred to as TM_0). This quantity thus includes effects of the initial conditions.
- $P_0(t)$ (referred to as P0(t0)), which is the probability of extinction by time t.

$P_0(t)$ is calculated in META-X using Eq. 1 with c_1 and T_m determined from the $\ln(1-P_0)$ plot. In addition, $P_0(t)$ may be influenced by a *relaxation time* which occurs if the initial state of the population was not worse but better than the established phase. For example, if in the established phase 5–10 of 15 patches of a metapopulation are occupied and the simulations are started with 15 occupied patches, it is almost certain that in every run the population will reach the established phase and consequently the probability of reaching the established phase, c_1, is one. However, it takes some relaxation time (Brooks et al. 1999) before the established phase is reached. This relaxation time can be estimated from $T_{mean}-T_m$ and is equal to the intercept of the linear fit of the $\ln(1-P_0)$ plot with the x- or time axis.

13.6 Dealing with Uncertainty

Fig. 13.3 is informative regarding the type of risk assessment which can be achieved with PVA. An *absolute risk assessment* would attempt to determine $P_0(t)$ with absolute precision. But such absolute assessments are virtually impossible because of the inherent uncertainty in model parameters and structure. Demographic and other vital parameters are usually not known precisely, and the model structure may be inappropriate because essential processes or structures of the population (e.g. density dependence) may be unknown.

Relative Risk Assessment

Instead, what most PVAs try to achieve are *relative risk assessments*, i.e. comparisons of the risk of extinction for different scenarios. If the relative errors of the risk assessments of the single scenarios are more or less the same, the insights from the comparison of the scenarios may still be valid. It is also important not to work statically with one single definition of viability, but to vary the tolerable risk (e.g. 1% or 5%) and the time horizon (e.g. 20, 50, or 100 years) in order to demonstrate that the classification 'viable' largely depends on our definition of viability.

Relative risk assessments are the main purpose of PVA because usually different management options are available, for example the provision of additional food in winter or additional breeding sites in summer (Drechsler et al. 1998,

Drechsler 2000). Which of these options would have the strongest positive effect on viability, or, if the PVA is about a negative impact, e.g. road construction, which alternative route of the road would have the weakest negative effect on viability? PVA would thus try to *rank* the management options.

Sensitivity Analysis

The basic idea of relative risk assessment is not to assess risk *per se*, but to assess how the risk of extinction *changes* if we change parameters, model assumptions, or management measures. *Sensitivity analysis* of model results is part of this basic idea: how does model output change if one of the parameters is changed while all the other parameters remain constant (i.e. have their reference values). In *local sensitivity analysis*, one parameter after the other is changed by a small amount, say 5%. Sensitivity is then quantified by the ratio of the relative change of the output variable, for example T_m, and the relative change of the parameter.

Low sensitivity means that the model results are robust (i.e. show little variation in response to changes in the parameters) and therefore uncertainty in parameter values may be ignored. High sensitivity, however, means that uncertainties of the corresponding parameters are highly relevant and the resources available should be focused on measuring these sensitive parameters, or – if possible – on formulating submodels with higher resolution.

Global sensitivity analysis means varying one parameter over its biologically meaningful range and recording the change of the output variable (for example Fig. 13.3). Such analyses give insights into functional relationships. However, these relationships may change if another reference parameter set were chosen. A full sensitivity analysis, where all possible parameter combinations are checked, is impossible for even just five parameters, say, let alone for more. Nevertheless, developing efficient strategies of sensitivity analysis is important and a focus of ongoing research (e.g. McCarthy et al. 1995, Drechsler et al. 1998).

13.7 Overview of PVA Models

So far, we have used an overly simple example model to demonstrate the basic concepts of stochastic population models and of viability analysis. Here, we give a very brief overview of PVA in the real world and the models which have been developed for PVA. For detailed reviews see Boyce (1992), Beissinger and Westphal (1998), Groom and Pascual (1998), Sjögren-Gulve and Hanski (2000) and Menges (2000).

The example model is a "Monte Carlo simulation model" because computer-generated random numbers are used to simulate demographic and environmental noise. The overall structure of such models for real PVA is just the same as in our example: nested loops over the runs, the years, and the individuals. If the model takes into account space explicitly, a loop over all the available habitats or all the subunits of the landscapes is added and usually arranged between the years and the individual's loop. In such spatially explicit models, dispersal is an additional process to be modelled.

The difference from the example model is that demographic and other processes are usually described in more detail. Depending on the information available, the life cycle of the individuals, their properties (e.g. age, sex, size, social rank) and their behaviour may be included. This means that birth and death rates are not mere numbers but are generated by flexible mechanisms. This has two advantages: first, the mechanisms might introduce a significant biological realism. For example, if a few individuals are buffered from the effect of environmental noise, this could significantly reduce extinction risk. And second, adequate resolution allows the model to be validated (Grimm et al. 1996). However, if the focus is on PVA of metapopulations, local population dynamics may may bescribed in a more aggregated way, as is the case with the META-X model (Chap. 14).

Monte Carlo simulations, however, are not the only way to model stochastic population dynamics. For very simple, unstructured models, analytical approaches can be used to derive analytical expressions for T_m. Usually, the 'diffusion-approximation' is used, and also more explicitly the theory of Markov processes. These models are not necessarily evaluated analytically but numerically (Wissel et al. 1994, Possingham et al. 2000). Often, the purpose of such theoretical models is to gain general insights and not to assess the extinction risk of a certain species. Therefore, the part of Theoretical Population Ecology which deals with such models and questions may be referred to as 'extinction theory'.

A rather new approach used in the past decade (Possingham et al. 2000) is to fit census time series to the stochastic versions of very simple population models, e.g. logistic or Ricker models. The model parameters gained from this fit are then used to analyse, for example, the Ricker model in just the same way as we analysed the example model.

Despite the bewildering diversity of stochastic population models used in PVA and extinction theory, the framework for evaluating these models, i.e. viability and intrinsic mean time to extinction, is the same (Grimm and Wissel 2002).

13.8 Potentials and Limitations of PVA

We have now given a brief introduction into the goals, methods and concepts of PVA. But what can really be achieved with PVA? PVA is now widely used in conservation biology; and while Boyce (1992) in his review noted that most of the PVA of that time was not accessible in publications, nowadays PVA is an established part of the conservation and ecological literature. However, parallel to this increased usage of PVA, criticism of PVA has also increased. This has three main reasons: first, as we mentioned in the introduction to this chapter, PVA is a tool and therefore, like any tool, can be abused. Abusing PVA is the consequence of being unaware of the goals, methods and concepts, and being uncritical regarding the potentials and limitations of PVA. Naturally, abusing a tool does not necessarily discredit the tool itself.

Second, the inherent uncertainty of data in PVA propagates uncertainties in the risk assessments achieved by PVA. Therefore, static or absolute risk assessment is no longer viewed as the purpose of PVA. Instead, PVA is being increasingly viewed as a 'decision support tool to help make management decisions' (Possingham et al. 2000).

And third, the primary predictions of PVA models, i.e. the prediction of persistence and viability, cannot be tested directly (but see Brook et al. 2000). But should management decisions be based on untested predictions? The focus on relative predictions partly solves this problem because they focus on mechanisms and understanding, not on mere numbers. In addition, modelling strategies which produce 'structurally realistic' models ("pattern-oriented modelling, Grimm et al. 1996, Wiegand et al. 2002) should be used because they have the advantage that they can be tested for the patterns which were used to design the model, and often also for other patterns. 'Secondary predictions' (Beissinger and Westphal 1998) concerning such additional patterns are indispensable for the credibility of the models used in PVA. Secondary predictions may concern, for example:

– characteristics of the census time series (floors and ceilings, cycles, trends, distribution of population sizes, reaction to disturbance events);
– structures within the population (age or size structure, group size distribution in social species, territory size distribution);
– spatial structures (presence/absence pattern in an archipelago of habitat patches).

Other ways of qualitatively testing PVA predictions are comparison with similar species (Possingham et al. 2000) and biogeographic patterns (Shaffer 1981). Species which are similar to the rare species in question but which are more abundant and can therefore be used to test whether the model is able to capture the essence of the population dynamics. And biogeographic patterns of the species itself, or of similar species, may indicate the minimum size habitat which can maintain viable populations. Storch (1995) estimated in this way the minimum capacity of habitats for capercaillie in Central Europe at about 500 individuals, which was largely confirmed later on by a PVA (Grimm and Storch 2000).

13.9 Metapopulation Viability Analysis

PVA is the viability analysis of populations. But what is a population? Is it an isolated population living in a more or less continuous habitat, for example a population in a reserve? Or is it a subpopulation of a larger population, e.g. the population of the reserve is part of the population of the region surrounding the reserve (Soulé 1987)? Clearly, 'population' may mean different things in different contexts (and disciplines, e.g. biogeography, ecology, genetics) and what really constitutes a population in a specific PVA is usually not self-defined but has to be defined by those who perform the PVA.

A particularly important type of populations is 'metapopulations' which consist of two or more subpopulations. These subpopulations live on habitat patches in a landscape or matrix of non-habitat. 'Habitat' means a place where the species can live and reproduce whereas in 'non-habitat' individuals will only stay for short periods of their life (dispersal) and in particular will not reproduce. The basic idea of the metapopulation concept is that the whole set of (sub)populations may be viewed as a 'population of populations' (Levins 1970) which go extinct and recolonize just the same as individuals die and are born in a population. A population may persist even though all the individuals will die sooner or later, and analogously a metapopulation may persist even though all the subpopulations may die out sooner or later. Individuals reproduce and so counterbalance mortality, and subpopulations recolonize empty patches and so counterbalance local extinctions.

One necessary condition for the 'metapopulation' effect of regional persistence is that the extinctions occur more or less independently on different patches; otherwise, in the extreme case of total correlation, all subpopulations could go extinct more or less simultaneously and no recolonization could occur. To have independent extinctions, however, means that the population dynamics on the patches have to be more or less independent. The basic mechanism behind the persistence of metapopulations is thus the partial decoupling of local (subpopulation) and regional (metapopulation) dynamics.

The analogy of metapopulation dynamics to population dynamics is also reflected in the two types of 'noise' that are responsible for extinctions. The 'turnover' of the metapopulation, i.e. the dynamics of extinctions and recolonizations, makes the number of occupied patches fluctuate and is analogous to demographic noise in spatially unstructured populations. Likewise, the partial correlation of extinction events on different patches is analogous to environmental noise in populations. The reason for this partial correlation is that environmental fluctuations may affect different patches synchronously. Weather, for example, might partly synchronize the population dynamics and, in turn, extinctions on different patches.

The concepts and methods for assessing metapopulation viability are identical to those used in PVA of spatially unstructured populations. Population or metapopulation viability analysis are thus the same, except that different types of models are used which address different processes (dispersal and spatial correlations in metapopulations).

Since about 1990, the concept of metapopulations has become almost a paradigm in conservation and population biology. One main reason for this is that due to anthropogenic fragmentation of most habitats, many populations are forced to assume a metapopulation structure: they live on habitat patches which are so small that their (sub)populations are not viable. The question, then, is whether the metapopulation is viable. If not, what could be done to achieve viability?

The success of the metapopulation concept also reflects the insight that ecology and, in particular, population ecology are basically spatial (Ranta et al. 1997, Blasius et al. 1999, Kendall et al. 2000). The change of locally observed variables, e.g. population size, is usually determined not only by local but also by regional processes. However, spatial dynamics and metapopulation dynamics are not necessarily the same. Care has to be taken not to infer metapopulation dynamics from a patchy structure of the landscape. The metapopulation effect of regional persistence may by superficial if one or more of the subpopulations are viable on their own, or if the coupling of the dynamics of the subpopulations via dispersal is so high that the local dynamics are no longer independent.

'Metapopulation' may be defined in different ways (e.g. Hanski 1991, Reich and Grimm 1996, Hanski and Simberloff 1997, Hanski 1999), but no matter which definition is used, for many so-called metapopulations the empirical evidence of the metapopulation effect of regional persistence is incomplete. For example, Reich and Grimm (1996) list the following four requirements of the metapopulation concept: (1) The subpopulations have their own dynamics which can be distinguished from the dynamics of other subpopulations and of the regional population. (2) At least one of the subpopulations is not viable. (3) Subpopulations interact via dispersing individuals. (4) Dispersing individuals are capable of colonizing empty patches, i.e. of establishing new subpopulations which by themselves produce colonizers. In a review of 87 empirical papers addressing 'metapopulations', Reich and Grimm (1996) found that in 64% of these papers the empirical evidence of one or more of the four requirements of their definition of metapopulation is lacking.

There can, of course, be no 'natural' definition of metapopulation and, as Hanski points out, "the task is not so much to classify species into one or another category ... . The real question is whether a *metapopulation approach* is useful or not. That is: are the assumptions valid that space is discrete; that ecological processes take place at two scales, local and metapopulation; and that the discrete spatial units of habitat are large and permanent enough to enable the persistence of local breeding populations for at least a few generations." (Hanski 1999, p. 3, emphasis in the original). Nevertheless, the caveat is to be aware of these implicit assumptions of the metapopulation concept; otherwise, attempts to understand and, ultimately, to manage a population for viability might be misguided.

13.10 Alternatives to Tailored PVA Models

PVA usually requires developing and analysing a (meta)population model. However, modelling and analysing models requires experience and a lot of time. Evidently, a new, tailored model cannot be developed for every threatened population. There are two major alternatives to tailored PVA models: rules of thumb drawn from extinction theory and specific PVA; and generic models which may be applied to many specific populations.

Rules of thumb are general guidelines for managing threatened populations. They refer to broad classes of ecological situations. For metapopulations, ecological situations are characterized by the properties of the species and the landscape involved. Consequently, the idea is to categorize species by a manageable set of 'profiles' (Grimm et al. 1996, Weaver et al. 1996, Frank and Wissel 1998, Vos et al. 2001) and the landscape by certain 'landscape indices' (With 1997, Wiegand et al. 1999). In general, rules of thumb put ecological situations into a hierarchical framework (Frank and Wissel 1998) and in this way allow management actions to be prioritized. For example, in an ensemble of five isolated patches there is no point in enabling dispersal among the patches unless the subpopulations on the patches have a certain minimum viability (Frank and Wissel 1998). The search for rules of thumb for managing threatened populations is promising but has not yet led to generally acknowledged results. Therefore, currently generic PVA models are the only applicable alternative to tailored models.

Generic models are designed to encompass as many specific situations as possible. They are distributed as ready-to-use software packages so that non-modellers and non-programmers can use them (for comparative overviews of generic models for PVA, see Lindenmayer et al. 1995, Brook et al. 1997, Brook et al. 2000a, b). The task of the user is to parameterize the generic model and to tailor it to some degree by choosing from among alternative modules. The design of generic models is based on the hierarchy of processes in populations and metapopulations. At the top of the hierarchy of population models are the age-specific birth and death rates of the individuals; they are the general currency of any population dynamics. However, data on these general rates often are not available and therefore tailored sub-models may be used to calculate these rates. Moreover, the demographic rates are usually not constant; tailored models could describe how they depend on biotic and abiotic factors. Generic models ignore all these detailed processes and focus on the general currency. Generic models compromise more detailed descriptions for general applicability (in META-X it is possible to include detailed processes via mechanistic submodels; Chaps. 14 and 15). They are thus between tailored PVA and no PVA at all, and the idea behind generic PVA models is that a compromised PVA is better than no PVA at all.

However, generic models have been criticized (e.g. Groom and Pascual 1998, Beissinger and Westphal 1998). The main target of criticism is that generic software packages may be too easy to use, i.e. by users who are unaware of the goals, concepts, methods and limitations of PVA. The purpose of PVA may become superficial if the generic packages are applied uncritically. For example, the strength of environmental noise largely determines extinction risk. There are, however, published applications of PVA packages where – because nothing was known about the strength of environmental noise – the default parameter proposed by the package was used. This may be a reasonable first step, but it has to be followed by a thorough sensitivity analysis – which was not performed! Thus, the blessing of PVA packages is also a curse: they are easy to use.

13.11 The Conception of META-X

META-X is not the first generic model and software for metapopulation viability analysis. VORTEX (Lacy et al. 1995, Lacy 2000) and RAMAS Metapop (Akçakaya 1997) are generic metapopulation models which have been maintained for at least five years and which are widely used. But META-X differs in some respects from these and other generic models:

1. META-X is a Levins-type metapopulation model, i.e. it does not explicitly consider the population dynamics of the subpopulations but only the two states of a patch 'empty' and 'occupied'. It has been shown theoretically (Drechsler and Wissel 1997) that such a coarse description of metapopulation dynamics is sufficient unless the time scales of subpopulation and metapopulation extinction are of the same order of magnitude. The virtue of using the Levins-type model structure is that the focus is on metapopulation dynamics. Details of local population dynamics are ignored. This keeps the model and its analysis simple.

2. The META-X model has a hierarchical structure (see Chap. 14): only the three sets of the main model parameters are needed to calculate persistence and viability. The main model parameters are extinction rates of the subpopulations, rates of colonization between each pair of patches, and the correlation of extinctions among each pair of patches. META-X uses several, partly alternative, submodels to calculate the rates of colonization and degrees of correlation. But the user may also use – and has to use in the case of local extinction rates – external submodels to generate the main model parameters. These parameter values may then be imported to META-X.

3. The conception of META-X emphasizes that PVA is comparative by its very nature. As already explained above, the purpose of PVA is not to produce individual numbers (i.e. single assessments of extinction risk) but to understand how extinction risk *changes* when model parameters are changed. The basic units to be simulated with META-X are thus not individual parameterizations of the model (i.e. scenarios) but comparative 'experiments' which consist of a suite of scenarios.

4. META-X is the first generic model whose assessment of extinction risk is based on the calculation of the 'intrinsic mean time to extinction', T_m. The most important advantage of this approach is that – once T_m has been determined – the extinction risk by time t, $P_0(t)$, can be calculated for any time horizon t of interest.

By these special features, we hope to utilize the advantages of generic models (no need to model those model structures which are generic) while avoiding the pitfalls of generic or 'canned' models. META-X 'cans' those elements of metapopulation viability analysis which may be canned, but leaves critical elements which require tailored submodels to the users. The idea is that beginners of PVA can use META-X to learn the goals, concepts and methods of PVA by applying the submodels of META-X to hypothetical scenarios. Then, step by step, users can learn to apply META-X to real populations and to use external submodels (see Chap. 15). At this stage, META-X might be coupled with other generic PVA models to calculate, for example, local extinction rates (for example POP-X, Köster et al. 2000, http://www.oesa.ufz.de/pop-x). Finally, users may develop their own submodels which are tailored to specific species and situations. We invite users of META-X to post links to their submodels which generate META-X parameters on the META-X homepage. The homepage might then help to integrate the efforts of numerous different metapopulation viability analyses and thus to overcome the 'PVA crisis', i.e. the lack of resources to develop tailored PVA models for every population of interest.

13.12 Suggested Reading

The book by Burgman et al. (1993) is a very useful introduction to the concepts and methods of risk assessment in conservation biology. A more recent primer on the technical concepts of PVA is given by Noon et al. (1998). The 'blue book' edited by Soulé (1987) is still a 'must' for those wanting to perform PVA because it was the starting point of the whole PVA story. In this historical context, Quammen (1996) is a nice, useful book to read: it describes the story and the people involved in the story of island biogeography and, in the last few chapters, of conservation biology and PVA. General introductions into conservation biology are given in Primack (1993) and Gibs et al. (1998).

Critical and very useful reviews of PVA are given by Boyce (1992), Caughley (1994), Doak and Mills (1994), Beissinger and Westphal (1998), Groom and Pascual (1998), Menges (1998, 2000; focusing on plant populations), and Possingham et al. (2000). A very useful review of the range of possible attitudes towards PVA is given by Burgman and Possingham (2000). Akcakaya and Sjögren-Gulve (2000) review the application of PVA in conservation planning.

Papers on the theoretical framework of PVA, i.e. extinction theory, are usually too advanced to be understood by beginners, but some basic introductions into extinction theory are given in the textbooks by Nisbet and Gurney (1982), Wissel (1989), Renshaw (1991) and Roughgarden (1998).

The metapopulation concept is covered in the volumes edited by Gilpin and Hanski (1991) and Hanski and Gilpin (1997), and the monograph on metapopulation ecology by Hanski (1999) is especially recommended. Verboom et al. (1993) give an excellent review of metapopulation models, while Reich and Grimm (1996) review theoretical and empirical metapopulation studies.

Generic models for PVA are critically assessed and compared in Lindemayer et al. (1995), Brook et al. (1997), Chapman et al. (2000), and Brook et al. (2000a, b).

Appendix: Technical Aspects of the ln(1-P_0) Plot

The data which are used for the ln(1-P_0) plot are generated by a stochastic process, i.e. a stochastic population model. Therefore, even if 1,000 simulations are run to determine T_m, it is not to be expected that T_m will exactly be the same if another 1,000 simulations are run (using different random numbers) and evaluated. However, the purpose of the ln(1-P_0) plot in particular, and of PVA in general, is not to determine a numerically exact value of T_m or $P_0(t)$. This would be unrealistic given the uncertainty of model parameters and assumptions. It therefore makes no sense to devote too much time and energy to optimizing the plot, i.e. the choice of the initial state, the number of simulation runs, and the time horizon. Variations of T_m by up to 5% or even 10% are not critical. Nevertheless, some rough guidelines can be given to produce fairly exact values of T_m.

We found that 1,000 runs over a time horizon of 1,000 years were nearly always sufficient. It is not necessary for the full distribution of the extinction time (Fig. 13.4) to be within the time horizon. Even if $P_0(t)$ only reaches rather small values within the time horizon (e.g. 0.2), the $\ln(1-P_0)$ plot still allows T_m to be determined with sufficient precision.

To avoid excessive errors in T_m, values of $P_0(t)$ larger than 0.97 are ignored in META-X. This is because large values of $P_0(t)$ are determined by the very few long extinction times in the 'tail' of the exponential distribution which – due to their low number – might considerably deviate from the underlying statistical distribution.

Finally it should be noted that grouping the extinction times of in a histogram with, for example, 20 classes is not essential for the $\ln(1-P_0)$ plot. The only purpose of this grouping is to allow a quick graphical assessment of whether the histogram exhibits the typical exponential decay. For the very $\ln(1-P_0)$ plot, however, no grouping of extinction times is needed and a class width of one should be used.

14 The META-X Model in Detail

14.1 Introduction

A generic metapopulation model which is to be used for many different species, landscapes and problems has to be general but still structurally realistic. Structural realism is crucial because users cannot change the model's overall structure, only tailor it to their problems by specific parameterizations. Structural realism means including the key structures and processes of metapopulation dynamics. In META-X, structural realism is achieved by focusing the main model on key structures and processes at the metapopulation level.

A metapopulation in META-X is a set of habitats or patches which may be either occupied by an established subpopulation or not. The META-X model is thus similar to the first metapopulation model developed by Levins (1969), which can be referred to as a winking patch (Verboom et al. 1993) or occupancy-type model (Sjögren-Gulve and Hanski 2000) because only the winking of the patches between occupied and empty, but not their detailed population dynamics, is considered. It has been shown theoretically (Drechsler and Wissel 1997) and in a case study on a metapopulation of badgers (Verboom et al. 1991) that the winking patch approach is sufficient to analyse metapopulation viability, as long as the extinction and recolonization rates of the patches are known.

Using a winking patch model as a generic model has two advantages. Firstly, the number of parameters to be directly entered for metapopulation viability analysis is kept small. This makes it simpler to use the model in teaching and for exploratory analyses. It also helps focus on the essentials of metapopulation dynamics and saves us from drowning in the details of the population dynamics of the individual subpopulations. Secondly, the winking patch approach acknowledges the hierarchical organization of metapopulations: rapidly changing dynamics at lower hierarchical levels (i.e. changing population sizes on the patches and dispersal of colonizers) may at the higher hierarchical level of the metapopulation be aggregated to constant parameters such as extinction and recolonization rates. This hierarchical organization in turn allows for a hierarchical parameterization of the model: users may either specify metapopulation level parameters (which in META-X are referred to as main model parameters) or lower level parameters of submodels describing, say, recolonization.

In contrast to Levins' original model, META-X is spatially explicit because numerous studies have shown that the spatial configuration of patches and the connectivity among the patches have a strong influence on a metapopulation's viability. 'Spatially explicit' means that the patches have an explicit position in a certain coordinate system and that the connectivity (all pairs of patches between which recolonization is possible) has to be specified explicitly. Moreover, in META-X the patches may differ regarding the parameters describing local population, for example local extinction risk. Another difference from Levins' model is that the META-X model is stochastic, i.e. chance events are taken into account.

14.2 The Main Model

In the META-X model, patch position is characterized by the coordinates of a certain point, usually the centre of gravity of the patch. Patch distances refer to distances between these centres, not to the distances between patch borders. In most cases, this idealization will have no strong effect on the analysis of metapopulation viability. If, however, effects of patch shape and size are suspected, model parameters describing recolonization and correlated extinction can easily be adjusted accordingly.

The META-X model is identical with that of Frank et al. (1994) and Frank and Wissel (1998). Technically speaking, the META-X model is a time-continuous Markov chain model for finite metapopulations. As is known from Markov chain theory (e.g. Nisbet and Gurney 1982; Honerkamp 1990), such models are completely determined by a set of possible states (here: all possible occupancy patterns) and so-called short-term transition probabilities, which describe the likelihood of transitions between different states within a certain short time step Δt.

The entire metapopulation consists of N patches, each of which is either occupied ($x_i=1$) or empty ($x_i=0$). The state of the whole metapopulation is thus given by the occupancy pattern $x=(x_1,...,x_N)$ and determined by the states of the individual patches. Within a sufficiently short time step Δt, a change in the occupancy pattern can only occur due to a single local extinction ($x_i: 1\rightarrow0$), correlated extinction of two subpopulations ($x_i, x_k : 1\rightarrow0$) or recolonization of an unoccupied patch ($x_j : 0\rightarrow1$). These three processes are described by corresponding short-term transition probabilities which are explained in the following section. The simulation algorithm which calculates the transitions is described in the Appendix to this chapter.

Local Extinction

If a certain patch i is currently occupied, there is a certain short-term probability $v_i\Delta t$ that the corresponding subpopulation will go extinct within time Δt due do within-patch population dynamics. The parameter v_i is given by the reciprocal value $v_i = 1/T_{1,i}$ of the intrinsic mean time to extinction $T_{1,i}$ of the local population in patch i. Since both $T_{1,i}$ and v_i merely summarize the local effects on subpopulation extinction, the parameter v_i is referred to as 'intrinsic local extinction rate'.

Correlated Extinction

Besides the within-patch dynamic factors, there may be additional, regional forces of local extinction. If certain patches i and j are currently occupied, there is a certain risk that both patches will go extinct simultaneously within the time step Δt by reasons of regional stochasticity (e.g. correlated weather conditions). If both local extinction processes are completely correlated, simultaneous extinction occurs with a short-term probability given by the geometric mean,

$$\sqrt{v_i \Delta t}\sqrt{v_j \Delta t} \, ,$$

of the local extinction probabilities. If both local extinction processes occur uncorrelatedly, however, this probability will be zero. Therefore, the short-term probability of correlated extinction is generally given by

$$c_{ij}\sqrt{v_i}\sqrt{v_j} \cdot \Delta t \, ,$$

where c_{ij} refers to the actual 'degree of correlation' between the patches i and j. When parameterizing META-X, the whole matrix of all pairwise correlations of local extinctions has to be specified. Considering only pairwise correlations is sufficient: in time-continuous Markov chain models the correlation between more than two patches is either zero or can be completely expressed in terms of all the pairwise correlations. Therefore, all the effects of regional stochasticity relevant for metapopulation viability are summarized in the matrix of pairwise correlations c_{ij} and the rates v_i of local extinction.

To obtain an intuitive understanding of the significance of correlated extinctions for metapopulation viability, note that with complete correlation, recolonization from occupied patches would be impossible because all patches would become empty at the same time. Correlation is a mechanism which limits metapopulation viability. It is therefore important to carefully consider mechanisms which correlate or decorrelate the extinction processes on different patches. It is particularly important to have an idea of the relationship between correlation and patch distance. META-X provides a simple submodel (which is described below) on this relationship, but you do not necessarily have to use this submodel; instead, c_{ij} can be specified directly.

Recolonization

If a certain patch j is currently empty, it can be recolonized within time Δt by individuals of all patches i which are connected to patch j and which are currently occupied. Therefore, the short-term probability of the recolonization of patch j is given by the sum

$$\sum_{i:x_i=1} b_{ij} \cdot \Delta t$$

taken over all patches i which are occupied. The term $b_{ij} \Delta t$ denotes the probability of patch j being colonized by patch i within Δt. In META-X, the parameter b_{ij} is referred to as the 'colonization rate' between patches i and j.

Note that 'colonize' does not only mean that individuals from patch i reach patch j but that they actually establish a new population on patch j, i.e. a population which is characterized by its intrinsic mean time to extinction, $T_{1,j}=1/\nu_j$. In general, b_{ij} will depend on patch distance, as well as on other factors such as the number of emigrants which are produced on patch i. As with correlation c_{ij}, the colonization rate b_{ij} can be determined from a META-X submodel which is described below, or from your own or other external submodels and then entered (or imported) directly, ignoring the META-X submodel.

14.3 The Submodels Provided by META-X

META-X provides simple standard submodels for calculating the probabilities of correlated extinctions, c_{ij}, and the probabilities of recolonization, b_{ij}. These models are referred to as 'standard' models because they are supposed to be useful in many (but certainly not all) cases.

A Submodel for the Degree of Correlation of Local Extinction

The correlation between the extinction processes on different patches may decrease with the distance between the patches (Moloney 1993). In the submodel for the degrees of correlation c_{ij} between two patches i and j, we assume an exponential decrease, i.e.

$$c_{ij} = \exp(-d_{ij} / d_0),$$

where d_{ij} is the distance between patch i and patch j measured from center to center. The submodel parameter d_0 is referred to as the 'mean correlation length'. For a patch distance d_{ij} equal to the mean correlation length d_0, the degree of correlation is e^{-1} or 36%. For $d_{ij} << d_0$, correlation is very strong and $c_{ij} \approx 1$, whereas for $d_{ij} << d_0$ correlation is negligible ($c_{ij} \approx 0$).

Note that although it generally seems reasonable to assume that the correlation of extinction processes decreases with distance, factors other than distance may be important as well. For example, consider habitats located at the south and north slopes of mountains (Dorndorf 1999). Weather in winter or summer may strongly correlate the population dynamics of all habitats in a region which have the same exposure, largely irrespective of their distance. On the other hand, populations on northerly and southerly exposed habitats may be influenced the most strongly by completely uncorrelated factors, e.g. weather in winter and in summer, even though they may be located side by side.

A Submodel for the Rate of Colonization

In the submodel for the rate b_{ij} with which patch i colonizes patch j, we take into account that a successful colonization is the result of three processes: the emigration of individuals from an occupied patch, dispersal to a target patch and, finally, immigration and the establishment of a new subpopulation on the target patch. The rate of colonization b_{ij} is thus modelled as

$$ b_{ij} = \begin{cases} \dfrac{E_i}{n_i} \cdot B_{ij} \cdot \dfrac{0.5}{I_j} & \text{if patch i and patch j are connected to each other} \\ 0 & \text{else} \end{cases} \tag{14.1} $$

where

E_i Mean number of emigrants leaving the occupied patch i per year.

n_i Number of connections from patch i to other patches.

B_{ij} Probability of an individual which started at patch i dispersing to patch j successfully reaching its target patch. In META-X, B_{ij} is referred to as 'reachability'.

I_j Number of immigrants needed on patch j to establish a new subpopulation with a 50% probability of success.

The following is assumed in Eq. (14.1): an occupied patch i 'emits' on average E_i emigrants per year. Each emigrant randomly chooses one of the n_i target patches, i.e. only those patches are potential targets which are connected to patch i. This means that the n_i target patches are assumed to have the same attractiveness. The probability of an emigrant successfully reaching a target patch j is given by the reachability B_{ij}. A submodel for B_{ij} is described below in the next section.

By assuming that each occupied patch i provides a certain pool of E_i emigrants which are uniformly 'allocated' over all the n_i connected patches, some sort of 'competition for emigrants' between the patches is modelled. This competition effect is a 'standard effect' in the context of metapopulation dynamics. It obviously occurs as a result of dispersal along corridors. However, this type of competition can also be found if the species possesses a dispersal strategy where the individuals actively move through a homogeneously perceived matrix and stay at the first possible patch (Heinz et al. 2002). The submodel for b_{ij} provided by META-X ($1/n_i$ in Eq. (14.1)) allows this competition effect to be described in a very simple way. This represents some progress because most colonization models such as the widely used incidence function model (Hanski 1994) neglect this competition effect.

The parameter I_j corresponds to the parameter describing successful establishment in Island Theory (MacArthur and Wilson 1967), as we will show in the following. In this theory, the probability $p_e(j,I)$ of a new subpopulation being successfully established on an empty patch j is described by:

$$p_e(j,I) = 1 - \exp(-y_j I) \approx y_j I \qquad (14.2)$$

where I denotes the number of immigrants into patch j and y_j being a patch-specific parameter which is the probability that one immigrant initiates a successful establishment. The approximation on the right-hand side of Eq. (14.2) only holds if the number of immigrants I is rather small. The parameter y_j can be expressed in terms of the number of immigrants I_j needed for a 50% (i.e. $p_e(j,I) \approx y_j I = 0.5$) successful establishment of a new subpopulation by inserting I_j in Eq. (14.2) which gives $y_j \approx 0.5/I_j$.

A Submodel for Reachability

META-X also provides a simple submodel for calculating the reachability B_{ij}:

$$B_{ij} = \exp(-d_{ij}/d_1), \qquad (14.3)$$

where d_{ij} is again the distance between patch i and patch j. The parameter d_1 is, analogously to d_0, referred to as the 'mean dispersal range'. For a patch distance of $d_{ij}=d_1$ the probability of a disperser crossing this distance is e^{-1} or 36%. This parameter represents a species-specific ecological attribute. Exponential approaches such as Eq. (14.3) are widely used to describe the functional relationship between the reachability and the distance between two patches.

14.4 Summary and Discussion

The submodels allow the main model parameters such as the degrees of correlation c_{ij} and the rates of colonization b_{ij} to be expressed in terms of spatial characteristics of the landscape (e.g. location of patches, net of connections) and relevant species' ecological attributes (e.g. correlation length d_0, mean dispersal range d_1). Figure 14.1 gives an overview of the main model and the standard submodels provided by META-X, and shows how these models are related to each other.

However, META-X does not provide a builtin submodel for calculating parameters describing the local populations, i.e. their rate of extinction, the number of emigrants they produce, or the number of immigrants needed to establish a new subpopulation on an empty patch. Thus, if META-X is to be used to support management decisions, external knowledge and submodels have to be used. META-X is thus not fully 'canned' (Grimm et al. 2002) but requires active users who can decide on additional tools to parameterize META-X (see Chap. 15). Likewise, if necessary the built-in submodels of META-X can also be overwritten with the results of external submodels. The only fixed structure of the META-X model is the main model with the main model parameters, v_i, c_{ij}, and b_{ij}.

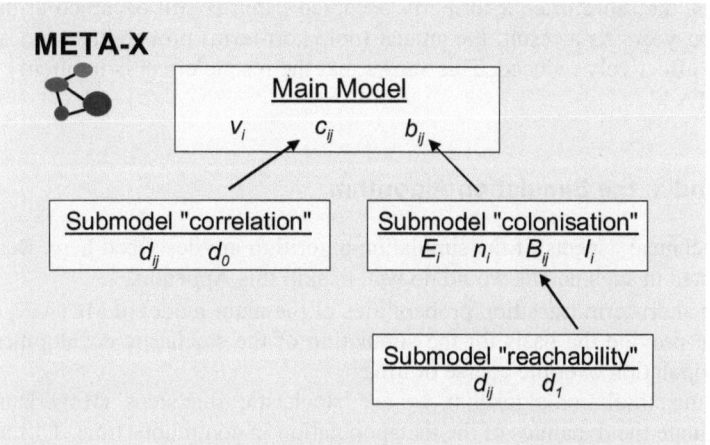

Fig. 14.1. Overview over the main model and the standard submodels provided by META-X.

Our decision to take into account in the model the correlation between local extinction processes (and, hence, the effect of regional stochasticity) led to a serious consequence: we were forced to choose a time-continuous modelling approach. In this case, the dynamics of the metapopulation is simulated as a sequence of events (e.g. single extinction, single recolonization) – in contrast to time-discrete models where all the events occurring in the course of one time step (e.g. a year) are simulated in a cumulative way (multiple events). Therefore, time-continuous models are usually more time-consuming than time-discrete ones. A longer computation time, especially for larger numbers of patches, is therefore the price for including the correlation.

Another important aspect related to the issue of correlation should be noted: an increase in the correlation leads to an increase in the effective rate of local extinction. This results from our assumption that a subpopulation can go extinct for two reasons: local within-patch dynamic effects or regional correlation-induced effects. The subdivision into local and correlated extinction, however, allows the local and regional factors of subpopulation extinction to be separately analysed in terms of their effect on metapopulation viability, i.e. to the benefit of a better understanding. There are alternative approaches to include the effect of regional stochasticity and spatial correlation (e.g. Burgman et al. 1993), each possessing methodological pros and cons.

META-X does not explicitly describe the rescue effect (Brown and Kodric-Brown 1977), which is the positive effect of immigrants augmenting the populations on occupied patches and thereby reducing local extinction risks. However, because of being time-continuous, the model automatically 'produces' the rescue effect. This can be seen by checking the metapopulation state produced by the model at the end of consecutive years. The typical phenomenon referred to is observed as a rescue effect: the stronger the exchange of individuals between the patches, the more often a formerly occupied patch is still occupied at the end of the next year. As a result, the annual (not short-term) probability of local extinction is effectively reduced. This shows that the rescue effect is implicitly included in META-X.

Appendix: the Simulation Algorithm

The technical aspects of the simulation algorithm are described here. Readers not interested in such details would do well to skip this Appendix.

The short-term transition probabilities of the main model of META-X, i.e. v_i, c_{ij}, and b_{ij}, provide the basis for the simulation of the stochastic development of the metapopulation over the course of time.

As the simulation algorithm, we use 'stochastic time steps' (Honerkamp 1990) to simulate the dynamics of the metapopulation in continuous time. This algorithm consists of two main steps. In the first step, it is determined how long the metapopulation will linger in its current state x or, in other words, how much time τ is required for a transition from x to another state x' to occur. In the second step, we determine which of the possible transition events ($x \rightarrow x'$) is realized. These two steps are described in more detail below.

Step 1: Determination of the (Stochastic) Transition Time

Assume that the current state of metapopulation is $x=(x_1,...,x_N)$. It is well known from Markov chain theory that the time τ until the next transition occurs is exponentially distributed and the corresponding exponential density distribution $f(t)$ is given by

$$f(t) = w_{tot}(x) \cdot e^{-w_{tot}(x) \cdot t} , \tag{A1}$$

where

$$w_{tot}(x) = \sum_{x_i=1} v_i + \sum_{x_i=x_k=1} c_{ik} \sqrt{v_i} \sqrt{v_k} + \sum_{x_j=0} \sum_{x_i=1} b_{ij} \tag{A2}$$

The term $w_{tot}(x)$ in Eq. (A2) is the total rate of possible transitions (local or correlated extinction, recolonization) from x into another state. Then, the transition time is determined in the following way:

- Draw a random number τ from the exponential distribution given in Eq. (A1).
- Determine the transition time t_1 by updating the current time t_0, i.e. $t_1 = t_0 + \tau$.

Step 2: Determination of the Transition Event

Which of the possible transition events from state x actually occurs at the transition time t_1 is randomly determined according to the following probabilities:

- $v_i / w_{tot}(x)$: extinction of the currently occupied patch i.

- $(c_{ik}\sqrt{v_i}\sqrt{v_k})/w_{tot}(x)$: correlated extinction of the currently occupied patches i, j.

- $(\sum_{i:x_i=1}b_{ij})/w_{tot}(x)$: recolonization of the currently empty patch j.

By repeatedly applying the described simulation procedure, we obtain a sequence of both consecutive transition times t_i and the metapopulation states $x(t_i)$ resulting from these transitions.

15 Parameterizing META-X

15.1 Introduction

One main task when working with META-X is to translate questions regarding hypothetical or real metapopulations into parameterizations of the META-X model. To be able to do this, you need to know the model and exactly what its parameters mean (Chap. 14). The next step is to compile all the relevant empirical information available and to extract the model parameters. To give you an idea of how to do this, we briefly describe some general concepts of parameterizing META-X (or any other PVA model) in this chapter. However, this is not the place for a complete introduction into the problem of parameterizing (meta-)population models. There is a whole body of literature on, for example, obtaining demographic or dispersal data from mark/recapture studies (Moilanen et al. 1998, Henle et al 1999, Moilanen 1999, Hanski et al. 2000). In general, if you want to learn how to parameterize PVA models, you will need to scan the PVA literature (see suggested readings at the end of Chap. 13) and assemble your own tool chest.

15.2 General Concepts of Parameterization

Direct Parameterization of Main Model Parameters

The most obvious way of obtaining the main model parameters of META-X would be to draw them directly from data. For example, the local extinction rate v_i could in principle be determined from observations. The snag is that, apart from extremely short-lived species studied in the laboratory, this would entail observing the metapopulation for hundreds of years. All the main model parameters of META-X basically deal with probabilities. In order to determine probabilities you have to observe quite a few events of extinction, recolonization or correlated extinction. This generally makes directly parameterizing META-X at the level of the main model parameters unfeasible.

Indirect Parameterization

As an alternative to direct parameterization, the idea behind indirect parameterization is that the information needed to determine model parameters may be hidden in certain observable patterns ('pattern-oriented' parameterization; Wiegand et al. 2002). We can only describe the general idea of indirect parameterization here; for technical details, please, consult the references given in this section. Consider, for example, a network of small and large patches; some of the patches are highly connected, others are rather isolated. If you now look at the occupancy pattern in a certain year, you can assume that – if the metapopulation is in a kind of stochastic equilibrium – large or connected patches are more likely to be occupied than small or isolated ones. The idea now is to construct a structurally realistic metapopulation model of this network and to fit the model output to the empirically observed snapshot (Hanski 1994). 'Fitting' here means adjusting the model parameters. Then, the best fit provides an indirect parameterization which is based on the occupancy pattern (Hanski 1994; 1999; an example application is given in Chap. 16).

Similar to occupancy patterns at the metapopulation level, patterns in time series can be used to parameterize demographic submodels. Wiegand et al. (1998) use time series of a brown bear population and of a certain environmental index to parameterize their demographic brown bear model; they were able to demonstrate that this indirect method avoids the problem of 'error propagation', i.e. the amplification of model uncertainty due to uncertainty in model parameters at lower hierarchical levels. Another field of research where the details of patterns in census time series are used is cyclic population dynamics. Here, the patterns are used to select among different model variants (Kendall et al. 1999).

Indirect parameterization is a rather new approach still being developed. For parameterizing META-X, this means that you should, apart from occupancy patterns and time series, also be aware of any kind of pattern that might, in a 'coded' way, contain information about the model parameters you are interested in. Such patterns might include age and size structures, group size distributions in social species (Dorndorf 1999; Grimm et al. 2002) or the frequency of certain events within a population (Stephens at al. 2002).

Mechanistic Submodels

Another alternative to direct parameterization of the main model parameters is to use mechanistic submodels to calculate these parameters. The submodels usually operate at faster time scales than the main model parameters. Therefore, shorter study periods are sufficient to obtain adequate data. The submodels include the standard submodels provided by META-X and user-defined, external submodels.

The submodels may partly be parameterized directly. For example, a demographic submodel of v_1 may be based on birth and death rates which can be determined from mark/recapture studies. Some species have even been studied in such detail and for such a long time that the birth and death rates can be calculated using for instance logistic regressions which quantify the influence of biotic and abiotic factors (Dorndorf 1999, Letcher et al. 1998). Alternatively, indirect parameterizations may also be tried for the submodels (see section above).

'Soft' Data, Empirical Knowledge and Hypotheses

Much of the frustration of those attempting PVA as well as those generally critical of PVA is based on the misconception that PVA has to be based entirely on 'hard' data, i.e. each parameter value has to be based on a large amount of numbers determined in the field. However, although hard data are desirable, they are usually only a subset of the empirical knowledge available. Empirical researchers who study a population for years or natural resource managers (e.g. foresters) often know much more than they could back up with hard data. For example, although the BEFORE model of natural European beech forests is based entirely on empirical rules which are not yet supported by hard data, it proved to be structurally realistic and was able to deliver independent, correct quantitative predictions (Neuert 1999; Neuert et al. 2001; Rademacher et al. 2001).

Focusing exclusively on hard data would bear the risk of ignoring empirical knowledge which may be based on decades of experience. On the other hand, the disadvantage of empirical knowledge (or 'soft' data) is that it may simply be wrong, irrelevant or based on prejudices. Therefore, soft data are uncertain and techniques to deal with uncertainty have to be applied. Then again, the same is – in principle – the case with hard data!

To obtain soft data, you have to interview those who know the system or species in question well. The interview should focus on information which is relevant to model parameters. For example, parameterizing a demographic model of the rock partridge (Stephan et al. 1995) entailed determining the strength of environmental noise (year-to-year variability of the population's growth rate). No hard data were available, but the empirical researchers and managers of the national park involved were able to answer questions such as: Which season causes more variation, summer or winter (winter)? How would you characterize the variability of winter strength (five classes) – and how would you assess the survival rates in each of these five classes? etc.

Another interviewing technique is to show results of draft models or arbitrary parameterizations to empiricists and to ask whether they consider these results as realistic. Time series, for instance, might be regarded as containing too much or too little variance or an unrealistic trend; likewise, population structure or spatial distribution may be unrealistic. Drechsler (2000) used this technique to rule out unrealistic parameter combinations.

Suppose there are no experts who know the system well or even the experts have no idea. In this case you have evidently identified an important gap in the empirical knowledge. If your entire parameterization consists of such gaps, you are indeed in trouble. But do not give up; try and see how far you can get with *hypotheses*. Hypothetical parameter assessments are based on biologically plausible assumptions. This may help you to at least narrow down the range of the parameter; for instance you might find arguments why mortality should be between 0.5 and 0.7, instead of between 0 and 1. Then, while analyzing the model, it may turn out that the range [0.5, 0.7] produces the full range of possible results, i.e. 100% probability of extinction or eternal persistence; in this case the PVA could indeed not be used to support management decisions. Alternatively, it could turn out that the ranking of management scenarios you are going to assess using your PVA is only marginally affected by the uncertainty of the parameter. Hypothetical parameter assessments are scientifically sound provided they are both clearly marked as being hypothetical and are taken into account when analyzing the model.

15.3 Dealing with Uncertainty

Uncertainty is inherent to PVA. Even the longest and most detailed field studies deliver parameters which are still uncertain to some degree. Moreover, PVA usually has to be based on very short, very coarse field studies (if at all), on soft data or even on hypothetical parameter assessments. Therefore, dealing with uncertainty is part and parcel of PVA. Below, we give some ideas of how to take uncertainty into account during parameterization. The general approach of sensitivity analysis is discussed in Chapter 13.

Upper and Lower Boundaries

For every model parameter at least three values should be provided: the reference value and upper and lower boundaries. The reference value may be the average value of field data or the 'best guess' of experts. The full set of reference values of all the parameters constitutes the reference parameter set which comprises the parameterization believed to best describe the situation given all the hard and soft data available. Upper and lower boundaries of the parameters describe the thresholds beyond which values are considered unrealistic or very unlikely (in general, upper and lower boundaries are well known as confidence limits). During the sensitivity analysis, the upper and lower boundary values for each parameter are used. This gives an initial idea of how important the uncertainty of the parameters is. However, this 'importance' is always linked to the reference parameter set used and may be different for other reference sets or, technically speaking, in other regions of the parameter space. It is also possible to analyze all possible combinations of upper, reference and lower values of all parameters (Drechsler 2000), although this will obviously lead to a very high number of parameter combinations, which may be hard to deal with.

Scenarios

The idea of upper and lower boundaries is to choose extremes for individual parameters. Similarly, one might go beyond individual parameters and try to specify extreme situations or scenarios, of the metapopulation in question. A pessimistic scenario, for instance, would reflect a pessimistic notion of the situation. Note that a pessimistic scenario is not necessarily the same as the parameterization where all parameters are assigned their lower boundaries. Instead, there may well be arguments why it is unlikely for all the parameters to reach their extreme pessimistic value at the same time – perhaps for example because the parameters are not independent. META-X is designed to support the definition and comparative analysis of scenarios.

Parameter Variation

PVA beginners often believe that parameterization and model analysis are two separate tasks to be performed one after the other. In fact, this is not a good strategy. Assume, for example, you find out that five of the parameters you are trying to determine are very uncertain. Now, you could carry on and try to reduce the uncertainty of all five parameters by, say, additional field studies, additional submodels, indirect parameterizations, etc. But then, after analyzing the model, three of the five parameters turn out to be largely irrelevant to the model results (e.g. the ranking of management options). This would be a waste of resources.

Therefore, parameterization and analyzing the model should be viewed as an iterative process. Parameterization means thinking in terms of ranges of parameter values, not individual numbers. META-X provides a convenient tool for dealing with such ranges: variation experiments (Chap. 11).

15.4 Some Specific Guidelines

Parameterizing the Landscape

In the landscape of META-X, the coordinates of the patch position correspond to the patch's centre or centre of gravity. The metapopulation is then visualized as a network of circles around the patch positions. These circles may represent local patch characteristics, e.g. local extinction rate, and do not indicate the spatial extension of the patches. Patch distance is measured from patch centre to patch centre. The matrices describing the degrees of correlation and colonization rates among all the pairs of connected patches are based on these patch distances. Although in many cases this will be sufficient, for certain species and landscapes it may be necessary to describe the patch shape and distance differently. To this end, users can modify or completely overwrite the matrices. This enables specific features of the landscape to be taken into account, i.e. features of the patch shape, barriers and corridors inhibiting or facilitating dispersal among patches or correlation of local extinction risk which is due to factors other than mere patch distance (e.g. features of habitat quality such as exposure).

Local Patch Characteristics

META-X provides no built-in submodel for the local patch characteristics, i.e. local extinction risk, the number of emigrants produced or the number of immigrants needed to establish, with 50% probability, a new subpopulation on the patch. These parameters generally require the use of external submodels. For the extinction rates, any generic PVA model of unstructured populations can be used, including POP-X (Köster et al. 2000, http://www.oesa.ufz.de/pop-x) because all these models deliver as a result the (intrinsic) mean time to extinction, which is the inverse of the extinction rate.

One submodel for the local extinction rate v_i that is widely used assumes a power-like dependence on the patch area A_i, i.e.

$$v_i = \varepsilon \cdot A_i^{-x}$$

with prefactor ε being some extinction parameter and power x describing the strength of environmental noise that summarizes the strength of environmental fluctuations and the species' sensitivity to them. The stronger these fluctuations and the more sensitive the species, the lower the value of x. This qualitative functional relationship between local extinction risk v_i and patch area A_i is well known and has been confirmed by numerous viability analyses for (both hypothetical and real) individual populations (e.g. Goodman 1987, Wissel et al. 1994, Foley 1994). It is also used in Hanski's practical Incidence Function Model (Hanski 1994).

To assess the average number of emigrants omitted by an occupied patch and the number of immigrants needed, a population model will usually have to be used – for example one of the existing generic PVA models of unstructured populations.

Correlated Extinctions

The degree of correlation, c_{ij}, determines the probability with which extinctions on pairs of patches occur simultaneously. If, for example, two patches are very close to each other and of similar habitat quality, a sequence of 'bad' environmental conditions will probably affect both subpopulations simultaneously. Consequently, both populations are likely to go extinct at the same time and therefore c_{ij} ≈ 1. On the other hand, if the distance between the two patches is large or if habitat quality is different, the degree of correlation can assumed to be small. The degree of correlation also depends on the spatial scale of the regional processes causing correlation. An epidemic, for instance, might influence the entire region in more or less the same way, whereas fire may affect only a subset of the patches. Basing parameterizations of c_{ij} on numerical data may usually be difficult. Therefore, the knowledge about habitat quality, regional processes and the main sources of fluctuations in the subpopulations' growth rates has to be translated into rough assessments of c_{ij}. Alternatively, a submodel for c_{ij} can be employed.

The standard submodel of META-X for c_{ij} assumes that the degree of correlation decreases exponentially with patch distance d_{ij}. The submodel parameter d_0 is referred to as 'mean correlation length': for a patch distance d_{ij} equal to the mean correlation length d_0, the degree of correlation is e^{-1} or 36%.

Note that although it generally seems reasonable to assume that the correlation of extinction processes decreases with distance, factors other than distance may be important as well. For example, consider habitats located on the southern and northern slopes of mountains (Dorndorf 1999). Weather in winter or summer may strongly correlate population dynamics of all the habitats in a region which have the same exposure, largely irrespective of their distance. On the other hand, populations on northerly and southerly exposed habitats may be influenced the strongest by completely uncorrelated factors, e.g. weather in winter and in summer, even though they even may be located side by side.

Colonization

The colonization rates b_{ij} can be parameterized directly or indirectly by using mark/recapture data or mechanistic submodels. There are different possible submodels for the colonization rates b_{ij}, each being applicable to a specific type of species dispersal. The submodel provided by META-X given by

$$b_{ij} = \begin{cases} \dfrac{0.5}{I_j} \cdot E_i \cdot \dfrac{1}{n_i} \cdot \exp(-d_{ij}/d_1) & \text{patch } i \text{ and patch } j \text{ are connected} \\ \\ 0 & \text{else,} \end{cases}$$

(Chap. 14; E_i: average number of emigrants; I_j: number of immigrants needed to establish a subpopulation with a probability of 50%) assumes that the individuals are strongly oriented towards connecting elements such as corridors adjacent to the starting patch during their dispersal, with d_1 being the mean distance an individual is able to move along a connecting element.

Another submodel is that implemented in Hanski's Incidence Function Model (Hanski 1994). It has the following, even more simple functional structure:

$$b_{ij} = y \cdot E_i \cdot \exp(-d_{ij}/d_1)$$

with y being some parameter describing the colonization pressure of a successful immigrant. Modelling the colonization rate thus is useful if a species with passive dispersal in a homogenous matrix (in particular without barriers) is considered.

A third possible submodel for the colonization rate b_{ij} is given by:

$$b_{ij} = \frac{0.5}{I_j} \cdot E_i \cdot \frac{R_{ij}^{N-1}}{\Sigma_{k(\neq i)} R_{ik}^{N-1}} \cdot R_{ij}$$

where

$$R_{ij} = 1 - \exp(-a \cdot e^{-d_{ij}/d_1})$$

with N being the number of patches and a being some parameter. This model is appropriate for species whose individuals actively move through a homogeneous matrix and immigrate into the first patch they arrive at. This 'immigration behaviour' produces some 'competition for migrants' between the patches that is expressed in the weighting factor given by the ratio of the R_{ij}^{N-1}'s. The predictive power of this submodel has been shown for a wide range of hypothetical as well as realistic movement patterns (Heinz et al. 2002).

Since META-X enables values for the colonization rates b_{ij} to be input from other files, any other submodel can also be used to parameterize the colonization rates.

15.5 Summary and Discussion

Burgman and Possingham (2000) describe a widespread, naive notion of PVA as "a tool that will deliver answers after an afternoon's playing with the computer" (p. 103). Once you have tried to parameterize META-X for your first real metapopulation or management problem, you will notice that neither an afternoon nor playing with a computer are sufficient for this task. For most PVAs performed with META-X, parameterization will require more time than working with META-X itself.

Parameterization takes hard work. Therefore, it is important to document this work because most of it will be devoted to dealing with uncertainties: "One of the most important steps in establishing the credibility of a PVA is to communicate the uncertainties embedded in the model and its assumptions." (Burgman and Possingham 2000, p. 104)

While trying to parameterize META-X, you will often get the feeling that it is virtually impossible to obtain sufficient empirical information to determine the parameters. But this uncertainty about model parameters is the rule in PVA, not the exception! This is why PVA does not deal with individual, fixed numbers (parameters, extinction risks), but with *ranges* of numbers. The real challenge of PVA and, in turn, META-X is not to overcome uncertainty but to deal with it (Chap. 13, Grimm et al. 2002).

Still, you may feel uneasy at the end of your PVA because uncertainties still exist. Then again, you made these uncertainties explicit. By contrast, conventional management and planning concerned with metapopulations, which does not include PVAs, is beset by exactly the same uncertainties but frequently ignores them! Burgman and Possingham (2000) claim that "one of the most important steps in establishing the credibility of a PVA is to communicate the uncertainties embedded in the model and its assumptions" (p. 104). It is therefore crucial to acknowledge that documenting the uncertainties of the parameters used in META-X is an important and critical part of PVA.

Appendix

We plan to maintain a list of parameterizations of META-X on the META-X homepage (http://www.oesa.ufz.de/meta-x). This list will provide examples and ideas for specific techniques of parameterization. META-X users are kindly invited to send us their parameterizations (for details, see the homepage) or links to their own websites where the parameterizations are described. We thus hope to compile relevant know-how about parameterization and to prevent every user from having to reinvent the wheel.

16 Example Applications

16.1 Introduction

Performing PVA requires performing experiments on the computer. The goal of these experiments is to understand how a metapopulation's viability depends on landscape structure and population processes. Therefore the particular submodels and parameterizations used should most appropriately be viewed as hypotheses or sets of assumptions about how things work in reality (Starfield 1997). The computer experiments then show the consequences of these assumptions.

Working successfully with META-X (or with any ecological model) requires adopting the attitude of experimenters (Grimm 1999, 2002): the starting point of a project always is a certain problem or question. Your task then, as an experimenter, is to design experiments which help answer the question or solve the problem. Sadly, there is no general protocol of how to translate problems into experiments. It is all a matter of learning by doing. To demonstrate how this translation works in principle, in this chapter we have compiled several projects which are also installed in the META-X directory on your computer (subdirectory 'Examples').

The examples are taken from the three areas where META-X is designed to be used: teaching and learning general metapopulation theory, analysing specific populations, and decision support in conservation and planning. The purpose of the examples is not to present in-depth analyses of the problems addressed but to demonstrate how META-X can, in principle, be used to tackle such problems. You are kindly invited to load the project files of the examples into META-X and to modify the experiments presented in this chapter depending on your own questions and hypotheses.

16.2 Examples of Theoretical Problems

Although every real metapopulation is unique, general principles still exist of how attributes of the species and the landscape affect metapopulation viability, e.g. the number of patches, their size and spatial configuration, and species attributes such as behaviour, life history and dispersal. The search for such general principles is an active area of research (e.g. Frank and Wissel 1998; Vos et al. 2001), and there is already something which may be referred to as 'metapopulation theory'. When performing metapopulation viability analysis and using META-X, it is important to be aware of this theory because theoretical insights make up the framework for posing appropriate questions, designing appropriate computer experiments, and making the right management decisions.

This is not the place to introduce metapopulation theory (for an overview, see Hanski and Gilpin 1997; Hanski 1999). What we want to do is to demonstrate that META-X can be used to learn metapopulation theory by doing, i.e. by designing experiments which address theoretical questions. The theoretical examples also show that theoretical insights may be directly relevant to decisions in conservation and landscape management.

Demonstrating and Analysing the Minimal Dispersal Range

Any metapopulation can be viewed as an assemblage of scales linked to the elements of the metapopulation (Stelter et al. 1997). Spatial scales are defined by, for example, the patch distances, the mean dispersal range of the species in question and the correlation of extinction risks on different patches. Temporal scales are defined by certain attributes of the species which, in turn, determine local extinc-

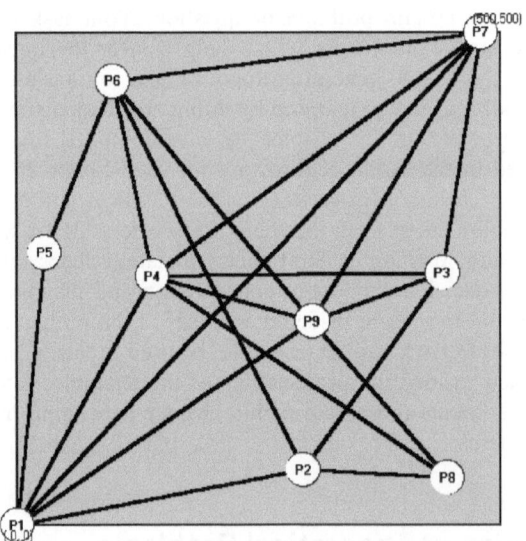

Fig. 16.1. Visualization of the arbitrary network of patches used for the theoretical examples. The patch symbols only indicate the position of the patches, not any local patch characteristics. The connections between the patches indicate pairs of patches between which recolonization is possible (after Grimm et al. 2002). The main model parameters are: v_i=0.4, d_0 =150 m. The size of the entire area is 500x500 m²

tion rates and recolonization dynamics. Viability, i.e. the ability of a species to persist in a certain landscape in the long run, requires the scales involved in this 'orchestra of scales' to fit each other (Stelter et al. 1997). Landscape attributes filter those species attributes which are required for a population to persist in the landscape, and vice versa. The following experiment will demonstrate this filter effect and the 'resonance' of scales which is required for viability.

First experiment: varying dispersal range d_1

Using the Landscape Editor or the Experiment Wizard, we first create an arbitrary configuration of patches whose patch-specific parameters are assumed to be identical (Fig. 16.1). Both the correlation length d_0 and the dispersal range d_1 are set to zero. The resulting scenario is taken as a basis for a 'variation experiment' (Chap. 10) where the dispersal range d_1 is varied between 0 and 1000 m.

Figure 16.2 shows how the mean time to extinction, T_m, of the metapopulation depends on d_1. Sufficiently high values of T_m are only obtained if d_1 exceeds a certain minimum value $d_{1,min}$, indicated by the line in Fig. 16.2. This means that a metapopulation can only persist in the long run in this landscape if the species in question has a certain minimum dispersal range $d_1 > d_{1,min}$.

The minimum dispersal distance is largely determined by the configuration of patches. A more detailed analysis shows that $d_{1,min}$ is approximately given by the arithmetic mean of the distances of the patches to their nearest neighbour, i.e.

$$d_{1,min} \approx \frac{1}{N} \sum_{i=1}^{N} \min\{d_{ij}, j \neq i\}, \tag{16.1}$$

where N is the number of patches and d_{ij} the distance between two patches i and j. This shows that the habitat network under consideration provides a spatial scale (right side of Eq. 16.1) to which the dispersal range d_1 (left side of Eq. 16.1) has to fit, otherwise no long-term persistence of the metapopulation is possible. Landscape management has thus to ensure that the patch configuration fulfills the criterion of Eq. (16.1).

Second experiment: varying correlation length d_0

The results of this first theoretical experiment lead to another question because we know that another spatial scale is decisive for metapopulation dynamics: the correlation length of local extinctions, d_0. To what extent will the relationship between the intrinsic mean time to extinction and the dispersal range d_1 be influenced by changes in d_0? In order to answer this question, we repeat the experiment described above for different values of d_0.

The different curves in Fig. 16.2 show the effect of increasing correlation length d_0. As long as d_0 is below a certain critical value $d_{0,crit}$, there is almost no effect on the relationship between T_m and d_1: the corresponding curves coincide with the curve belonging to $d_0=0$. Above the critical value, however, a change in d_0 leads to a characteristic change: while the minimum dispersal range $d_{1,min}$ remains

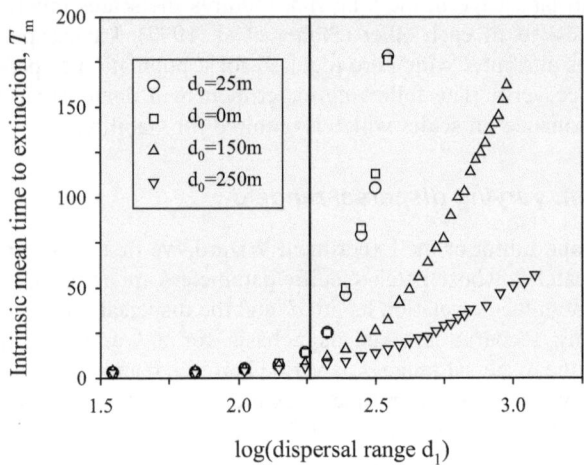

Fig. 16.2. Intrinsic mean time to extinction, T_m [years], versus logarithm of mean dispersal range, d_1 [m], for different mean correlation lengths, d_0, for the network of patches of Fig. 16.1. The vertical line at $\log(d_1)=2.26$ corresponds to $d_1=181$m which is the mean distance to the nearest neighbour patch

the same, the increase in T_m with d_1 is slower for larger d_0. This shows that a large dispersal range d_1 is not sufficient to compensate for the negative effect of a large correlation length d_0.

This effect can be explained as follows: as long as d_0 is below the critical value, the distance d_{ij} between neighbouring patches is large enough to ensure that no patch is within the range of correlation of any other patch. Consequently, the metapopulation effectively behaves as a metapopulation with correlation length $d_0=0$. The situation changes when d_0 exceeds the critical value $d_{0,crit}$. Now the range of correlation around a certain patch becomes larger. More and more patches are within the correlation range, which increases the risk of a successful recolonization of an empty patch being counteracted by simultaneous extinctions of several local populations. As a result, the relative importance of successful colonizations and, hence, of the species' dispersal range decreases.

Discussion and lessons

Particularly the latter findings have the following three serious implications for conservation management: (1) A change in the spatial scale of critical impacts (enlargement of the correlation length d_0) can bring a formerly safe metapopulation to the edge of overall extinction (if d_0 exceeds $d_{0,crit}$). This shows that correct conclusions regarding the viability of a metapopulation can only be drawn if the correlation length is taken into account. (2) If the correlation length is already above the critical value $d_{0,crit}$, it makes no sense to focus exclusively on an improvement in the conditions for species' dispersal. Instead, effort should be invested in asynchronizing or stabilizing the local dynamics. (3) It can be shown that, as a rule of thumb, the critical correlation length $d_{0,crit}$ is approximately given by a third of the arithmetic mean of the distances from the patches to their nearest neighbours, i.e.

$$d_{0,crit} \approx \frac{1}{3}\frac{1}{N}\sum_{i=1}^{N}\min\{d_{ij}, j \neq i\}$$ (16.2)

This shows that the critical correlation length is strongly determined by the spatial scale provided by the spatial configuration of the habitat network under consideration. To summarize, this very simple theoretical experiment does indeed corroborate the general notion that metapopulation viability depends on how the scales of the system fit each other.

What are the lessons of these insights for conservation and management? The assessment of metapopulation viability and the ranking of alternative management options requires thinking in terms of the scales of the system. Certainly, it is not easy to know the mean dispersal range and, in particular, the mean correlation length precisely, but note that in Fig. 16.2 the abcissa has a logarithmic scale. The question on the relationship of scales is thus in fact more a question of orders of magnitudes than a question of precise numerical values.

Now consider a real scenario where a number of small patches are more or less isolated from each other. Let us assume that enough is known about the species to predict that the isolated subpopulations will go extinct very soon. One management option is to establish links between the patches for recolonization (an idea which was, for example, so popular in Germany in the 1970s that linking habitat patches was considered the ultima ratio of virtually all conservation problems). However, our experiment shows that this management would almost certainly fail if the order of magnitude of dispersal range is smaller than the minimum dispersal range. Moreover, linking the patches may be much less effective than expected if the correlation length is larger than the critical correlation range $d_{0,crit}$ (Eq. 16.2).

In such cases, management would have to focus on ways to affect the correlation regime of the landscape. Although this sounds very theoretical, simple practical means exist which would affect correlation: managing habitat quality on the patches asynchronously (Stelter 1997), or trying to add patches of different quality to the network, etc.

Another scenario would be a loss of, on average, every second habitat patch in the landscape, for example due to increased land use in the region. What changes is not d_1 or d_0, but the mean patch distance. This warns us not to view any scale of the system as absolute, but rather as relative. In this hypothetical scenario, the mean dispersal range affects metapopulation viability via the mean patch distance. If the mean patch distance increases, a former uncritical dispersal range may become critical. The consequence for management is, again, that any assessment of impacts – deliberate or otherwise – on the system requires a critical consideration of the interplay between all the spatial and temporal scales involved.

Increasing the Number of Emigrants from a Certain Patch

In the above example, we considered a homogeneous landscape, i.e. all patches had the same characteristics or, technically speaking, the same parameters (local extinction risk, number of emigrants, number of immigrants needed). Of course, considerations of metapopulation persistence become much more complicated and case-specific if we consider heterogeneous landscapes. However, introducing all kinds of heterogeneity into an experiment at the same time and asking unspecifically how heterogeneity affects viability would be a poor research strategy. A better bet would be to perform 'controlled' experiments where individual factors are changed one at a time. After all, controlled experiments are the basis of all predictive, 'hard' natural sciences such as physics, chemistry or molecular genetics.

Below we consider the theoretical problem of how metapopulation viability changes if on a certain patch the mean number of emigrants produced per year is increased.

Experiment: increasing the number of emigrants

For this experiment, we use the same landscape as in the experiment above (Fig. 16.1) and vary the number of emigrants, E, on patch 9. An increase in E leads to an increase in T_m (Fig. 16.3) up to a saturation value of T_m. Initially, T_m increases with increasing E because more colonizers increase the chance of empty patches being recolonized. Once E is large enough to guarantee that virtually all empty patches are recolonized immediately, a further increase in E has no effect on T_m. However, the positive effect of colonizers originating from patch 9 depends on the occupancy of this patch, i.e. on how likely it is that this patch itself is occupied by a subpopulation. This becomes evident if we perform the same experiment

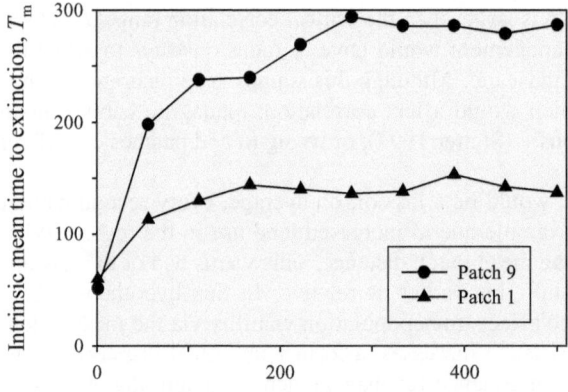

Number of emigrants, E, from a certain patch

Fig. 16.3. Intrinsic mean time to extinction, T_m, versus the number of emigrants, E, from patches 1 and 9, respectively

as before, but now vary E on patch 1, which has the same parameters as patch 9, but instead of being at the centre is located at the periphery of the habitat network and therefore has a smaller incidence. The increase in T_m with E is now less marked and the saturation value of T_m lower than in the case of patch 9 (Fig. 16.3).

Discussion and lessons

This example demonstrates how subtle the effects of changing patch characteristics may be in a heterogeneous network. Even if management only affected one specific patch, it would be impossible to predict the effect of this management by focusing on just this patch. Instead, the whole network, i.e. the specific configuration of patches, their size and their connectivity, determine the significance of certain patches for the viability of the entire metapopulation. Metapopulation dynamics is largely determined by this reinforcement of negative or positive effects of patch characteristics and configurations. Simple, linear cause–effect reasoning is inappropriate for understanding and managing metapopulations.

The lesson from this example for conservation is that exclusively managing individual patches may have not the desired effects if the integration of the patch into the network of patches is not appropriately considered. If possible, management should always address the entire network instead of single patches.

16.3 Examples of PVA for Specific Metapopulations

When using META-X to assess and understand the viability of real metapopulations, two tasks are required of the user: translating the problem or question into an experiment, and parameterizing the META-X model according to the real system. Although parameterizing will be the more time-consuming and difficult task, note that the experimenter's attitude is still needed. Even the best data sets available will not allow parameters to be specified with absolute certainty. Particularly in conservation, most data sets are inherently poor because the species in question was not or cannot be studied in more detail. Parameterizing real species thus does not mean mapping reality into one single parameter set because this is simply impossible. Instead, there will be a reference parameter set which is assumed to be the best possible approximation of the real system. The robustness of the results obtained with this reference parameter set must then be checked by carefully designed sensitivity analyses where one parameter is changed while all the others are left unchanged (for the rationale of sensitivity analysis, see Chaps. 13 and 15).

Besides sensitivity analysis, the experimenter's attitude is also decisive if we take into account the 'time axis' of any scientific or applied project concerning real metapopulations. It would be ineffective to first collect data for many years and then to use, for example, META-X to integrate all these data and to test working hypotheses about the system analysed. Instead, incomplete and uncertain assessments of the system parameters could be initially analysed with META-X, taking into account the uncertainty by performing many comparative experiments and analysing many alternative scenarios and hypotheses. The results of these preliminary analyses will very probably enable the following steps of research to be focused on the main questions, processes or parameters. Thus, using META-X 'on the fly' is likely to make research more effective.

The same holds for applied problems. For example, the first models on the Spotted Owl demonstrated that exactly those aspects of this species which were virtually unknown (such as juvenile mortality during dispersal) were the most sensitive parameters for assessing area requirements (Noon and McKelvey 1996). In turn, these initial models were later criticized, especially when they were used as the basis of real management decisions. However, since the first PVA model on the Spotted Owl was published, much research has been conducted to build up our knowledge of this species. It is questionable whether this research would have been performed without the early PVA models on this species. Thus, uncertainty or even a lack of data should not prevent PVA (Starfield 1997); on the contrary, they should stimulate PVA in order to concisely demonstrate the state of the art of the knowledge base regarding a certain species.

Below we give two examples of how META-X is applied to specific metapopulations. These examples will also demonstrate how to use the hierarchical structure of the META-X model by using external submodels. In the first example, a full external metapopulation model is used for parameterization. In the second example, a separate submodel for local population dynamics is used. The parameters produced by the external submodels are then imported into META-X.

Ecological Uncertainty and the Ranking of Management Options: the Butterfly *Melitaea diamina*

The metapopulation of this example consists of 17 habitat patches of the butterfly *Melitaea diamina* which are part of a larger habitat network in Aland, Finland (Fig. 16.4). This metapopulation has been thoroughly studied by Hanski and co-workers. We use it below as an example for parameterizing META-X using presence/absence data and the incidence function model (IFM) by Hanski (1994, 1999).

External submodels

To parameterize META-X for this problem, we directly use the parameterization of Hanski's IFM (1994) calculated by Wahlberg et al. (1996). The IFM is very similar in form to the META-X model. To be exact, it is a winking-patch model and thus ignores details of local population dynamics and dispersal. The important thing is that the IFM can be parameterized by using the presence/absence data of just one single snapshot of the system, assuming that the metapopulation is in the established state. The idea is to use regression techniques which compare the presence/absence data with the incidence of the patches in the model. Regression delivers a parameter set of the IFM which produces realistic occupancy patterns. For a detailed description of the IFM approach, see Hanski (1994, 1999) and Moilanen

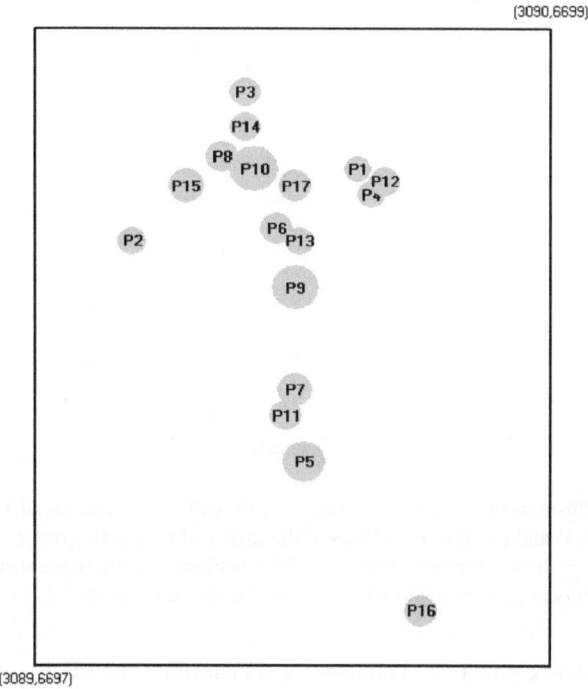

Fig. 16.4. Visualization of the network of 17 habitat patches of the butterfly *Melitaea diamina* in Aland, Finland (after Wahlberg et al. 1996). All patches are, in principle, connected and so the connections not shown. The size of the patch symbols indicates the intrinsic mean time to extinction of the corresponding population (the 'largest' patch having a T_m of about 4 years; for our demonstration example, the original patch sizes have been reduced by 95%).

et al. (1998). User-friendly computer tools to calculate IFM parameters from presence/absence data are available on the internet (see, for example, the links listed on the META-X homepage).

The relationship between the parameters of the META-X model and the IFM submodels is:

$$v_i = e\,A_i^{-x}$$
$$b_{ij} = g\,A_i^{b}\,\exp(-d_{ij}\,\alpha)$$
$$c_{ij} = 0.$$

The meaning of these relationships is as follows. The local extinction risk v_i is calculated from a very simple submodel which uses patch size A_i as a surrogate of population size and two parameters, e and x. The probability of recolonization between two patches i and j, b_{ij}, decreases exponentially with the patch distance d_{ij}, and depends on the number of emigrants, which is assumed to depend on the

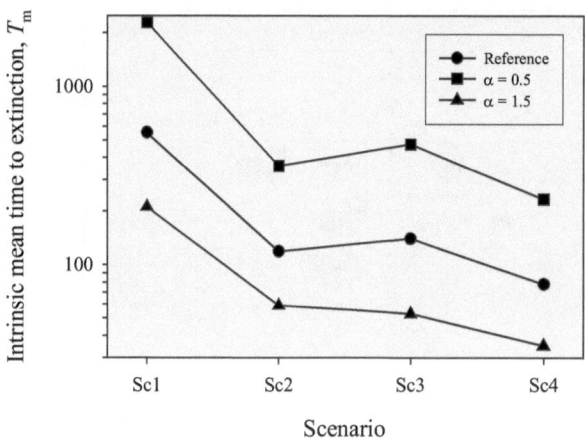

Fig. 16.5. Intrinsic mean time to extinction, T_m, for different scenarios of the butterfly example (Fig. 16.4) and for different values of the dispersal parameter (α=0.5: dispersal range doubled; α=1.5: dispersal range reduced by 2/3). Scenarios: *1* reference scenario; *2* patch 5 is lost completely; *3* patches 9 and 10 lose 50% of their area; *4* patch 9 is lost completely

area of the source patch, A_i. Parameter g transforms connectivity to colonization probability, parameter α defines the effect of distance on migration, and parameter b gives the scaling of emigration as a function of patch area. The IFM model typically ignores possible correlations of extinctions on different patches (c_{ij}=0), although correlation in dynamics can be added to the model (see e.g. Moilanen et al. 1998).

Using the IFM and presence/absence data, Wahlberg et al. (1996) determined the following set of parameters: x=0.884, e=0.014, g=3.62, b=0.5, α=1.0. We used these values and the IFM submodels to calculate the main model parameters for the metapopulation presented in Fig. 16.4. These parameter values were then imported to META-X (see the META-X homepage for details).

First experiment: ranking management options

The management question we are going to discuss is hypothetical and will be used to demonstrate how the uncertainty of model parameters can be taken into account. The scenarios we are going to compare are:

1. The original configuration.
2. A large patch at the periphery of the network (patch 5) is lost.
3. Half the area of two patches (patches 9 and 10) is lost.
4. A large patch from the centre of the network (patch 9) is lost.

The total area lost is assumed to be the same in all three habitat loss scenarios. The question now is: what is the ranking of the three alternatives with respect to the metapopulation persistence, and: how robust is the ranking with respect to uncertainties in specific ecological parameters of *Melitaea diamina*? Figure 16.5 shows the result of the META-X experiment. The ranking of the scenarios according to their negative effect on the metapopulations' persistence is: 3, 2, and 4, i.e. scenario 3 has the highest persistence.

However, results obtained by META-X (or any other ecological model) should never be blindly accepted (for example, the occurrence of low-level mistakes during parameterization can never be ruled out). Instead, always critically discuss numerical results and try to find a verbal explanation. In this experiment, the relative ranking of scenarios 2 and 4 is obvious: patches at the periphery contribute less to recolonization than central patches. However, why is the loss of a complete peripheral patch more critical than reducing the area of two central patches by 50% (scenarios 3 and 4)? The reason is that even a reduced central patch is still able to play its 'role' in the network dynamics. Moreover, the complete loss of the peripheral patch reduces the total number of patches and thereby increases the risk of metapopulation extinction due to the stochastic turnover dynamics (the metapopulation equivalent of demographic noise in small populations).

Certainly, all these results may depend on the specific patch configuration and species-specific parameters. Feel free to design additional 'if-then' scenarios and exploratory experiments of your own!

Second experiment: sensitivity of ranking to uncertainty

As an example of uncertainty in model parameters, let us assume that we are not sure about parameter α, the parameter describing the dispersal range (α corresponds to $1/d_1$ in the META-X model). Therefore, we perform the same experiment as in Fig. 16.5, but use $\alpha=0.5$ and $\alpha=1.5$, which corresponds to doubling and reducing the mean dispersal range by 2/3, respectively. The results of these experiments are presented in Fig. 16.5. Of course, changing α affects persistence as quantified by T_m. This comes as no big surprise since we know already (see experiment with varying dispersal range above) that the dispersal range and, in turn, recolonization have a strong influence on the persistence of the metapopulation. However, the ranking determined with the reference parameter set does not or only slightly change (scenarios 2 and 3 change their ranking orders in case of $\alpha=1.5$; but note that the difference in the corresponding T_m-values is only small).

Discussion and lessons

The uncertainty of model parameters is indeed a severe problem for PVA if absolute assessments of viability are required. Then again, absolute assessments are not the purpose of PVA, because they are simply impossible. Ecology has long been known as not being terribly successful at numerical predictions of certain ecological variables. Why should this be suddenly different for PVA, which is based on stochastic ecological modelling of small populations?

The lesson from our example is that as long as the uncertainty concerning model parameters does not amount to orders of magnitude, relative assessments of viability may be rather robust. If, under certain circumstances, they are *not* robust, this would indicate specific crucial thresholds in the metapopulations' dynamics which we need to be aware of.

Metapopulation Scenarios of Capercaillie *Tetrao urogallus* in the Bavarian Alps

In this example (Grimm et al. 2002) we look at the capercaillie (*Tetrao urogallus*), a grouse occurring in boreal or montane forests in Eurasia. Many populations in Central Europe have gone extinct during the past five decades (Storch 2001) and in Germany, only the populations in the Alps and the Black Forest are considered viable, i.e. with an extinction risk of less than 1% in 100 years (Grimm and Storch 2000). In the Bavarian Alps, mountain forests still seem to offer an appropriate habitat for the capercaillie. However, most mountain forests are rather small and separated from each other by unsuitable habitat in the valleys. Most populations in a single mountain forest patch consist of 100 or less individuals, which is below the minimum requirement for a capacity of about 500 individuals (Grimm and Storch 2000). The distance between neighbouring mountain forest patches is mostly 5–10 km, a distance over which juvenile capercaillies can disperse (Storch and Segelbacher 2000). It has therefore been hypothesized that the capercaillie populations in the Bavarian Alps constitute a metapopulation (Storch 1993; Storch and Segelbacher 2000). To test whether existing empirical knowledge on the demography and dispersal of the capercaillie supports the metapopulation hypothesis, META-X can be used.

Parameterization

Parameterization involves several steps. Each step is only a rough approximation of reality because the goal is to obtain a rough, initial idea of the hypothetical metapopulation dynamics. The steps are:

1. Delineating the patches. This is roughly achieved by using a GIS to identify those forested areas which are higher than 1000 m above sealevel. In the following, we consider only a subset of the patches in the Bavarian Alps (Fig. 16.6). Note that patch 86 is actually much larger than assumed here, the majority of the patch being in the Austrian Alps.

Table 16.1. Parameterization of a network of local populations of capercaillie in the Bavarian Alps

No.	X	Y	Area	Radius	K	T_m	ν	I	E
39	10.00	16.08	9.55	1.743	19	24	0.041	4	5
47	22.30	17.52	14.93	2.180	29	42	0.023	5	8
48	18.40	20.37	54.34	4.158	108	230	0.004	12	20
76	31.07	14.02	37.62	3.460	75	121	0.008	10	15
86	20.65	10.00	110.94	5.942	221	919	0.001	24	40
57	29.96	19.47	11.42	1.906	22	30	0.033	5	8
71	38.78	22.00	19.39	2.484	38	54	0.018	6	10
70	35.04	26.96	12.62	2.004	25	34	0.029	5	8

X, Y: Patch coordinates [km]. K: Patch capacity [individuals]. *Area*: [km²]. T_m: Mean time to extinction years]. ν: Local extinction risk per year. I: Number of immigrants needed for 50% probability of establishment of new population. E: Average number of emigrants produced by a patch. Patches: 19: Steilenberg/Rauhe Nadel; 47: Gurnwandkopf; 48: Hochgern/Hochfelln; 76: Sonntagshorn; 86: Seegatterl/Dürnbachholm; 57: Rauschberg; 70: Teisenberg; 71: Hochstaufen.

2. The patches are characterized by the coordinates of their centre of gravity and by their area. The area is converted to a ceiling capacity of the patches by assuming an average capacity of two individuals per square kilometre.
3. The resulting capacities are used as parameters in a demographic model of the capercaillie parameterized for the Bavarian Alps (Grimm and Storch 2000). The model delivers the extinction rates of the subpopulations.
4. The number of emigrants produced by the patches was estimated based on population size, age structure and patch size. The number of immigrants needed to establish a new population on an empty patch with a probability of 50% was determined with the demographic model and using the $\ln(1-P_0)$ plot (see Chap. 13). The plot allows the probability of establishment, c_1, to be determined.
5. The model parameter 'mean dispersal range', d_0, was estimated for female juveniles based on dispersal data of Koivisto (1963), i.e. $d_0=10$ km, although these data were obtained in the contiguous forests of Finland, not in fragmented landscapes. Nevertheless, these are the best data on capercaillie dispersal available. The 'mean correlation length' c_0 was assumed to be zero.

Experiments

For the patches in Fig. 16.6a and the parameters specified above and in Table 16.1, the metapopulation is viable, i.e. it has an extinction risk smaller than 1% over 100 years. The incidence, i.e. the probability of being occupied, of all patches is close to 1 (Fig. 16.6b). Viability would even be maintained if, for example,

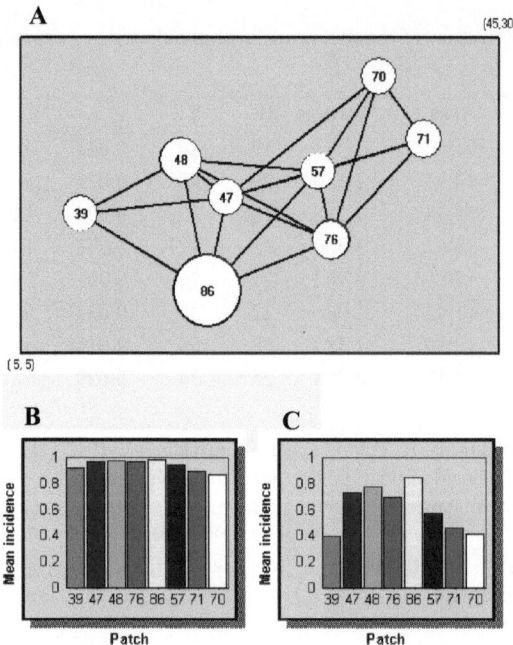

Fig. 16.6. <u>A</u> Visualization of the landscape used in the capercaillie example. Patch size indicates the local intrinsic mean time to extinction, ranging from 919 years (patch 86) to 24 yrs (patch 39; Table 16.1). <u>B</u> Incidence (= probability of being occupied by a population) of the patches for the base scenario. <u>C</u> Incidence for a scenario where d_0 = 4 km (after Grimm et al. 2002).

patches 39, 47 and 57 did not exist. However, the dispersal range may have been overestimated because dispersing birds might tend to avoid open areas. The metapopulation is still viable for d_0=5 km, but no longer for d_0=4 km (T_m=4,900 years, which corresponds to an extinction risk of roughly 2% in 100 years). In this scenario, incidence is much smaller, and three patches are occupied only with an incidence of less than 50% (Fig. 16.6c).

Discussion and lessons

These results support the notion that the numerous small populations in the Bavarian Alps are integrated within a viable metapopulation, or even a spatially structured population, and that under current conditions, i.e. with no further habitat deterioration, no loss of incidence is to be expected – unlike elsewhere in Germany, where most populations have gone extinct or seem doomed to extinction. However, even this initial exploratory analysis shows that the amount and range of dispersal among patches is critical. Since only anecdotal data are available on the dispersal of capercaillie in fragmented landscapes, other dispersal submodels than the META-X submodel should be tested. Likewise, other elements of the model such as the number of patches considered, local capacities, the numbers of emigrants and immigrants, dispersal barriers, and the specific pattern of environmental correlation should be tested regarding their significance for viability. Since the existing data on demography and dispersal are not sufficient to parameterize a detailed PVA model which would produce testable predictions on population time series or structure, the purpose of such analysis can only be to assess the relative importance of processes and structures. Further questions that could be addressed regarding the capercaillie metapopulation in the Bavarian Alps are whether there is a pessimistic but realistic scenario which could endanger the metapopulation, for example, an increase in nest predation, increasing fragmentation, increasing isolation, or habitat deterioration and is the view of this network of small populations as a metapopulation supported by genetic analyses? (Storch and Segelbacher 2000).

16.4 Example of Decision Support for Landscape Management

In landscape management, the purpose of PVA is to support rational decisions, i.e. decisions which are based on rational consideration of all the information available. Below we consider by way of example a hypothetical problem which has in common with many other applied problems the fact that a decision has already been taken to do *something* which will decrease the viability of a certain metapopulation: a highway is to be constructed which will cross a network of habitat patches of a certain, threatened species (a similar problem is discussed in Grimm et al. 2002). The question now is as follows. There are alternative routes for the highway which would cut through the network at the periphery, at the centre, or in between these two extremes (Fig. 16.7). To mitigate the situation, in each of these three scenarios, a 'green bridge', i.e. a bridge covered with soil and vegetation acting as a corridor, is planned to be constructed. Which of the scenarios minimizes the loss of viability of the metapopulation?

Experiments

To compare the effect of the three different routes of the highway, we make two assumptions. First, the highway will indeed sever the links between the adjacent patches on either side. Dispersers will still head towards the patch on the other side but will, it is assumed, be killed on the highway, i.e. the reachability, B_{ij}, is set to zero for the corresponding pairs of patches. Second, we assume the green bridge makes the movement safer than under natural conditions ($B_{ij}=0.8$ instead of 0.4 for the pair of patches linked by the green bridge as in the reference scenario). The persistence of the original, undisturbed situation and of the three alternative scenarios is presented in Fig. 16.8. The clear result of the comparison is that the closer the highway runs through the centre of the network, the worse it will be for

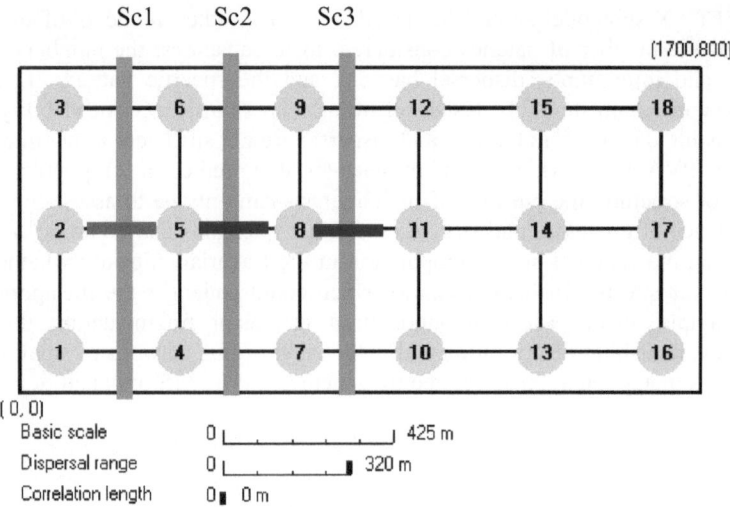

Fig. 16.7. The three scenarios of the highway example (see text for explanation).

the metapopulation. This finding is known from general considerations of the problem of fragmentation: the more peripheral the location of the isolating agent (in this case the highway), the larger the undisturbed remaining network and, in turn, the larger the persistence of the remaining network. In none of the three scenarios is the coupling of the two network fragments on either side of the highway strong enough to allow for a 'higher-order' metapopulation effect where each network fragment could conceptually be considered as a single patch.

The drawback of all three alternatives is that dispersers which try to cross the

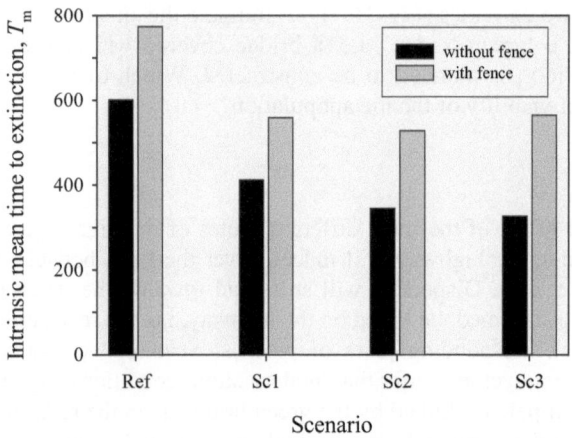

Fig. 16.8. Intrinsic mean time to extinction, T_m, for different scenarios of the highway example.

highway are simply lost (which may be a reasonable assumption for busy highways) and therefore cannot contribute to the persistence of the metapopulation. Therefore, in a new experiment we will explore the consequences of the following modified assumptions. First, we assume that a fence has been constructed which indeed cuts the connections between the patches on both sides of the highway. Note that this assumption means – due to the recolonization submodel of META-X – that the emigrants are now distributed among the remaining links to other patches and therefore are no longer lost to the system. Secondly, we assume that the effectiveness of the green bridge is the same as in the first experiment.

For scenarios 1 and 2, the results are, relative to the undisturbed scenario, similar to those of the first experiment (Fig. 16.8), but now scenario 3, where the highway passes right through the centre of the network, performs better than the other two scenarios. The reason for this effect is that now in scenario 3, the network on the left-hand side of the highway is large enough to establish a moderately viable metapopulation of its own; and since now the link with the metapopulation on the right-hand side of the highway is more effective, a meta-metapopulation effect does indeed emerge (i.e. the left-hand part can recolonize the right-hand part after an extinction). The disadvantage that, while moving from 1 to 3, the right network becomes smaller at a certain point is more than counterbalanced by the fact that a second viable metapopulation becomes possible.

This is an interesting, by no means obvious, effect of constructing the fence. But note that this effect will only work if the green bridge has a certain, minimal effectiveness (we leave it to you to design experiments where the effectiveness of the green bridge is varied).

Discussion and lessons

This example again demonstrates that simple, 'linear' reasoning may not be sufficient to assess the effect of alternative impacts on an existing metapopulation. As we mentioned earlier in this chapter: the effects of manipulating metapopulations are subtle and may be counterintuitive. In the real world, considerations like those presented in this example could help limited resources to be used more efficiently than if decisions were taken off the cuff. In reality, the three alternative routes of the highway will not cost the same in terms of either construction costs or socio-economic costs. On the other hand, a green bridge, a more effective green-bridge and a fence do not come free either, and it is important to demonstrate to decision-makers when and how and why they are effective.

16.5 Summary

The purpose of this chapter was to demonstrate how META-X can be used in the fields of teaching, analysing specific populations and decision support in conservation and planning. All the examples demonstrated that adopting the experimenter's attitude is crucial. We also tried to show that META-X is a tool to deal with, and not to overcome, uncertainty (Grimm et al. 2002).

However, META-X is no less rigorous or even serious than other generic PVA models. Not focusing too much on individual numerical values of parameters of extinction times does not mean that META-X is less 'scientific' or reliable. The reliability of any PVA in general, and of META-X in particular, depends on the quality of the data and empirical knowledge, and on the understanding of how a metapopulation's structures and processes interact and thereby determine viability.

We recommend that you modify the experiments presented here. Formulate your own questions, try to design appropriate experiments and try to understand the results. If you do not understand them, design additional experiments. You can try the same with real or theoretical metapopulations described in the literature: try to project these metapopulations onto META-X scenarios! META-X is designed for learning by doing and playing!

However, we would like to emphasize that, although the experiments presented here are easily designed and performed, thorough PVAs need a considerable amount of time and work. The naive notion of PVA which is described by Burgman and Possingham (2000) as "a tool that will deliver answers after an afternoon's playing with the computer" (p. 103) is certainly misguided. Parameterizing the model, sensitivity analyses, designing meaningful experiments and distilling and communicating the main results of a PVA require hard work (but may still be fun).

Glossary

(Re)Colonization

Occurs when a new subpopulation is established on an empty patch by immigrants from other, occupied patches. 'Established' means that the size of the subpopulation displays typical fluctuations linked to the constant risk of short-term extinction described by the local extinction rate ν_i.

Comparative Risk Assessment

The purpose of META-X or any other PVA model is to assess the risk of extinction, for example for different management scenarios. However, since in virtually all PVA the parameters are more or less uncertain, it is impossible to achieve absolute risk assessment, i.e. exact, certain numerical results. Instead, PVA is designed to compare risk assessments of different scenarios (including parameter variations), i.e. comparative (relative) risk assessments.

Connectivity

A pair of patches is said to be connected if mutual recolonization is possible. The connectivity of the network of patches has to be specified by the user.

Control Parameters

The two control parameters specified by the user, number of runs and time horizon, describe how often the simulation of a certain parameter set is repeated and the maximum number of years individual simulations may last.

Correlated Extinction

If the local extinctions on the different patches occur completely independently of each other, they are completely uncorrelated and the degree of correlation is zero. If, on the other hand, all subpopulations go extinct at the same time, the degree of correlation is equal to one. In most cases therefore, the degree of correlation will have a value between 0 and 1. The degree of correlation is determined by regional factors such as weather, epidemics or habitat quality.

(Mean) Correlation Length

Parameter quantifying the patch distance over which simultaneous extinctions on different patches are very likely to occur. This parameter is used in the standard submodel of META-X to calculate the degree of correlation on pairs of patches. In this submodel, the degree of correlation is assumed to decrease exponentially with patch distance, scaled by the parameter 'mean correlation length'.

(Mean) Dispersal Range

Parameter describing the patch distance over which emigrants are very likely to reach other patches. The parameter is used in the standard submodel of META-X to calculate the reachability between pairs of patches. This submodel assumes an exponential decrease of reachability with patch distance, scaled by the parameter 'mean dispersal range'.

Evaluation

The raw results of simulations are evaluated regarding certain quantities of interest, e.g. intrinsic mean time to extinction or incidence of patch occupancy. 'Evaluation' refers to both the process of evaluation and its results.

Experiment

Basic unit to be simulated and evaluated by META-X. An experiment consists of one or – usually – more scenarios. The basic idea of an experiment is that it is comparative because the purpose of the experiment is to compare different scenarios.

Extinction Risk

Probability of extinction within a certain time span, for example one year (short-term extinction risk) or a time horizon of interest, e.g. 50 years.

Homogeneous Parameters

For certain theoretical questions, it is convenient to ignore the explicit spatial configuration of the landscape or the variability among local patch characteristics on different patches. For such situations, all corresponding model parameters (e.g., local extinction rates, reachabilities) are set to the same (=homogeneous) value.

Import

In META-X several model parameters or groups of parameters, can be imported from external ASCII files. These parameters are patch positions, local patch characteristics, correlation matrix, reachability matrix and colonization matrix.

Incidence

Probability of a patch of being occupied by a subpopulation. Usually, large or highly connected patches have a higher incidence than small or isolated patches. In META-X, patch incidence is determined simply by dividing the number of years a patch is occupied by the total number of years for which it is observed. The incidence of patches is an indicator of the role a patch plays within the landscape: most important to metapopulation persistence are those patches with a high incidence. Incidence also indicates the importance of certain connectivity patterns.

Intrinsic Mean Time to Extinction

This is the mean (=average) time to extinction of a metapopulation which has reached the 'established phase' where the number of occupied patches undergoes typical fluctuation and the short-term risk of extinction is constant. The intrinsic mean time to extinction is determined using the '$\ln(1-P_0)$ plot'.

Landscape

A landscape in META-X consists of a certain clipping of the landscape and of the network of patches and connections between them.

Landscape Editor

A META-X tool to visualize the landscape of a certain scenario, to modify scenarios via a graphic interface and to create scenarios. The Landscape Editor is located in the right half of the META-X window.

Local Aspects

Patches are characterized by their positions and their parameters (patch characteristics). Each of the three patch characteristics can be chosen to scale the size of the symbols representing the patches in the Landscape Editor. If no local aspect is chosen, all patch symbols have the same size.

Local Extinction

Extinction of a subpopulation on a certain patch.

Main Model Parameters

The META-X model has a hierarchical structure. The main model describes the main processes of a metapopulation, viz. local extinction, recolonization and correlated extinction. Accordingly, the main model parameters are local extinction rates, the matrix of mutual colonization rates and the matrix of the degrees of correlation (of local extinction).

Metapopulation

A metapopulation consists of a set of (local) subpopulations living on discrete habitat islands (patches). The population dynamics of the subpopulations are not entirely correlated. Patches which became empty due to local extinction may be recolonized from occupied patches.

META-X Model

The metapopulation model underlying META-X.

Parameter

Ecological models such as META-X deal with variables and parameters. Variables describe the state of the system being modelled (e.g. number of occupied patches) and usually change with time. Parameters are numbers specified by the user and quantify functional relationships and processes, e.g. dispersal range or local extinction risk.

Parameterization

The process of translating empirical data, empirical knowledge or assumptions into parameters.

Patch

A discrete island of habitat suitable for living and reproduction surrounded by a matrix through which individuals may pass, but where they do not settle and reproduce.

Patch Characteristics

The three parameters describing a patch and the subpopulation that may live on it: local extinction rate, average number of emigrants emitted by the patch and average number of immigrants needed to establish a subpopulation if the patch is empty.

Persistence

Ability of a small (meta)population, which is strongly influenced by chance events, to persist, i.e. not to go extinct. Persistence is quantified by the time to extinction. Since this time varies due to stochastic processes, the entire distribution of extinction times has to be considered or alternatively the average of this distribution, i.e. the (intrinsic) mean time to extinction.

Project

A unit which contains one or more experiments and which is also the unit which is saved to (or loaded from) disk as a *.mtx file.

Project Tree

The tree diagram in the left half of the META-X window visualizes the elements of a project, i.e. experiments and scenarios. Many of the procedures used in META-X can be performed via the context menus linked to the elements of the Project Tree, such as starting simulations, displaying evaluations, copy&paste of scenarios or experiments, etc.

Reachability

The probability with which individuals starting from a certain patch will reach another patch.

Report

Reports are generated by the Report Generator of META-X. They contain the documentation of the parameterization and evaluation of the scenarios of a project.

Scenario

A full parameter set of the META-X model is known as a 'scenario'. Although scenarios may be simulated and evaluated individually, scenarios are usually compiled in an experiment and then simulated and evaluated comparatively.

Sensitivity Analysis

Since most model parameters are usually uncertain, it is important to vary all the parameters and check the resulting changes in model results. If the results are robust (i.e. do not change much), the uncertainty of a parameter can largely be ignored; if they are sensitive, whether the uncertainty will affect the ranking of management options needs to be assessed. Moreover, parameterization should focus on the most sensitive parameters.

Simulation

Running the META-X model on a computer means running a simulation. Simulations of scenarios may be performed interactively or automatically. The units of simulations are experiments or selected individual scenarios.

Stochastic Model

If a model includes a description of random variations, it is known as a stochastic model. In computer models, random processes are implemented using computer-generated random numbers.

Submodel

Mechanistic submodels are used to determine the main model parameters of the META-X model. 'Mechanistic' means that these models refer to the mechanisms underlying the main model parameters, whereas the main model parameters themselves only represent the results of these mechanisms. For correlated extinctions and for reachability, 'standard' submodels are provided by META-X. These standard submodels can be overwritten by external submodels – which are needed anyway to parameterize the patch characteristics. The results of external submodels may be imported while applying the Scenario Wizard.

Time Horizon

To limit the computation time needed for simulations, the control parameter 'time horizon' is specified. If a simulated metapopulation is still alive at the end of this horizon, the simulation is stopped and the next simulation started. Time horizons of 300 or 1,000 years are usually chosen. If the time horizon chosen is too short, too few extinctions may be recorded, which will not allow the persistence and viability of the metapopulation to be quantified. Note that 'time horizon' is a technical control parameter and is not directly related to the time interval of interest over which the viability of the metapopulation is to be assessed, e.g. 50 years.

Uncertainty

As a rule, the parameters of ecological models are not known precisely because too few data are available. The parameters are thus more or less uncertain.

User-Defined

If in a scenario some or all of the parameters produced by the standard submodels are overwritten by the user, the scenario is referred to as 'user-defined'.

(Parameter) Variation Experiment

An experiment consisting of scenarios in which all the parameters except one are the same. This individual parameter is varied incrementally between certain boundaries. Variation experiments are used for sensitivity analyses.

Viability

The ability to persist with a particular certitude (e.g. 95%) over a certain time interval (e.g. 100 years).

Wizard

A sequence of windows guiding the user to perform a certain sequence of tasks such as parameterizing a scenario.

References

Akcakaya HR (1997) RAMAS (R) Metapop: viability analysis for stage-structured meta-populations. Version 2. Applied Biomathematics, Setauket, New York, USA

Akcakaya HR, Sjögren-Gulve P (2000) Population viability analyses in conservation planning: an overview. Ecological Bulletins 48: 9-21

Beissinger SR, Westphal MI (1998) On the use of demographic models of population viability in endangered species management. J Wildlife Manage 62: 821-841

Blasius B, Huppert A, Stone L (1999) Complex dynamics and phase synchronization in spatially extended ecological systems. Nature 399: 354-359

Boyce MS (1992) Populaton viability analysis. Annu Rev Ecol Syst 23: 481-506

Brook BW, Lim L, Harden R, Frankham R (1997) Does population viability analysis software predict the behaviour of real populations? A retrospective study on the Lord Howe Island woodhen *Tricholimnas sylvestris* (Sclater). Biol Conserv 82: 119-128

Brook BW, Burgman MA, Frankham R (2000a) Differences and congruencies between PVA packages: the importance of sex ratio for predictions of extinction risk. Conservation Ecology 4: 6. [online] URL: http://www.consecol.-org/vol4/iss1/art6

Brook BW, O'Grady JJ, Chapman AP, Burgman MA, Akcakaya HR, Frankham R (2000b) Predictive accuracy of population viability analysis in conservation biology. Nature 404: 385-387

Brooks TM, Pimm SL, Oyugi JO (1999) Time lag between deforestation and bird extinction in tropical forest fragments. Conserv Biol 13: 1140-1150

Brown JH, Kodric-Brown A (1977) Turnover rates in insular biogeograhpy: effect of immigration on extinction. Ecology 58: 445-449

Burgman M, Possingham H (2000) Population viability analysis for conservation: the good, the bad and the undescribed. In: Young AG, Clarke GM (eds) Genetics, demography and viability of fragmented populations. Cambridge University Press, Cambridge, pp 97-112

Burgman MA, Ferson S, Akcakaya HR (1993) Risk assessment in conservation biology. Chapman & Hall, London

Caswell H (1989) Matrix population models. Sinauer, Sunderland

Caughley G (1994) Directions in conservation biology. J Anim Ecol 63: 215-244

Chapman AP, Brook BW, Clutton-Brock TH, Grenfell BT, Frankham R (2000) Population viability analyses on a cycling population: a cautionary tale. Biol Conserv 97: 61-69

Doak DF, Mills LS (1994) A useful role for theory in conservation. Ecology 75: 615-626

Dorndorf N (1999) Zur Populationsdynamik des Alpenmurmeltiers: Modellierung, Gefährdungsanalyse und Bedeutung des Sozialverhaltens für die Überlebensfähigkeit Doctoral thesis, Philipps-Universität Marburg,

Drechsler M (1998) Sensitivity analysis of complex models. Biol Conserv 86: 401-412

Drechsler M (2000) A model-based decision aid for species protection under uncertainty. Biol Conserv 94: 23-30

Drechsler M, Burgman M, Menkhorst PW (1998) Uncertainty in population dynamics and its consequences for the management of the orange-bellied parrot *Neophema Chrysogaster*. Biol Conserv 84: 269-281

Drechsler M, Wissel C (1997) Separability of local and regional dynamics in metapopulations. Theor Popul Biol 51: 9-21

Foley P (1994) Predicting extinction times from environmental stochasticity and carrying capacity. Conserv Biol 8: 124-137

Frank K, Drechsler M, Wissel C (1994) Überleben in fragmentierten Lebensräumen - Stochastische Modelle zu Metapopulationen. Z Ökologie u Naturschutz 3: 167-178

Frank K, Wissel C (1998) Spatial aspects of metapopulation survival: from model results to rules of thumb for landscape management. Landscape Ecol 13: 363-379

Frank K, Wissel C (2002) A formula for the mean lifetime of metapopulations in heterogeneous landscapes. Am Nat 159: 530-552

Gibbs JP, Hunter MLj, Sterling E (1998) Problem-solving in conservation biology and wildlife management: exercises for class, field, and laboratory. Blackwell Science, Malden

Gilpin M, Hanski I (1991) Metapopulation dynamics: empirical and theoretical investigations. Academic Press, San Diego

Goodman D (1987) The demography of chance extinction. In: Soulé ME (ed) Viable populations for conservation. Cambridge University Press, Cambridge, pp 11-34

Grimm V (1999) Ten years of individual-based modelling in ecology: what have we learned, and what could we learn in the future? Ecol Model 115: 129-148

Grimm V (2002) Visual debugging: a way of analyzing, understanding, and communicating bottom-up simulation models in ecology. Natural Resource Modelling 15:23-38

Grimm V, Frank K, Jeltsch F, Brandl R, Uchmanski J, Wissel C (1996) Pattern-oriented modelling in population ecology. Sci Total Environ 183: 151-166

Grimm V, Lorek H, Finke J, Koester F, Malachinski M, Sonnenschein M, Moilanen A, Storch I, Wissel C, Frank K (2002) META-X: a generic software for metapopulation viability analysis. Biodiv Conserv (in press).

Grimm V, Storch I (2000) Minimum viable population size of capercaillie *Tetrao urogallus*: results from a stochastic model. Wildlife Biology 5: 219-225

Grimm V, Wissel C (2002) The intrinsic mean time to extinction: a unifying approach to analyzing persistence and viability of populations. (submitted)

Groom MJ, Pascual MA (1998) The analysis of population persistence: an outlook on the practice of viability analysis. In: Fiedler PL, Kareiva PM (eds) Conservation biology: for the coming decade. Chapman & Hall, New York, pp 4-26

Hanski I (1991) Single-species metapopulation dynamics: concepts, models and observations. Biol J Linn Soc 42: 17-38

Hanski I (1994) A practical model of metapopulation dynamics. J Anim Ecol 63:151-162

Hanski I (1999) Metapopulation ecology. Oxford University Press, Oxford

Hanski I, Alho J, Moilanen A (2000) Estimating the parameters of survival and migration of individuals in metapopulations. Ecology 81: 239-251

Hanski I, Gilpin ME (1997) Metapopulation Biology. Ecolgy, Genetics, and Evolution. Academic Press, San Diego

Hanski I, Simberloff D (1997) The metapopulation approach, its history, conceptual domain and application to conservation. In: Hanski I, Gilpin ME (eds) Metapopulation biology. Academic Press, San Diego, pp 5-26

Heinz SK, Conradt L, Wissel C, Frank K (2002) Dispersal behaviour and fragmented landscapes: a formula for the probability of reaching a patch. (submitted)

Henle K, Vogel B, Köhler B, Settele J (1999) Erfassung und Analyse von Populationsparametern bei Tieren. - In: Amler K, Bahl A, Henle K, Kaule K, Poschlod P, Settele J (eds) Populationsbiologie in der Naturschutzpraxis. Isolation, Flächenbedarf und Biotopansprüche von Pflanzen und Tieren. - Ulmer, Stuttgart, pp 94-112

Honerkamp J (1990) Stochastische dynamische Systeme: Konzepte, numerische Methoden, Datenanalysen. VCH Verlagsgesellschaft, Weinheim

Kendall BE, Bjornstadt ON, Bascompte J, Keitt TH, Fagan WF (2000) Dispersal, environmental correlation, and spatial synchrony in population dynamics. Am Nat 155: 628-636

Kendall BE, Briggs CJ, Murdoch WW, Turchin P, Ellner SP, McCauley E, Nisbet RM, Wood SN (1999) Why do populations cycle? A synthesis of statistical and mechanistic modeling approaches. Ecology 80: 1789-1805

Koivisto I (1963) Über den Ortswechsel der Geschlechter beim Auerhuhn (*Tetrao urogallus*) nach Markierungsergebnissen. Die Vogelwarte 22: 75-79

Köster F, Stephan T, Finke J, Sonnenschein M (2000) Ein Simulationswerkzeug zum praktischen Einsatz in Naturschutz und Landschaftsplanung - ExiDlg. In: Möller DPF (ed) Frontiers in Simulation, Simulationstechnik 14. Symposium in Hamburg, ASIM, pp 503-508

Lacy RC (2000) Structure of the VORTEX simulation model for population viability analysis. Ecological Bulletins 48: 191-203

Lacy RC, Hughes KA, Miller PS (1995) VORTEX: A stochastic simulation of the extinction process. Version 7 user's manual. IUCN/SSC Conservation Breeding Specialist Group, Apple Valley, MN.

Letcher BH, Priddy JA, Walters JR, Crowder LB (1998) An individual-based, spatially-explicit simulation model of the population dynamics of the endangered red-cockaded woodpecker, *Picoides borealis*. Biol Conserv 86: 1-14

Levins R (1969) Some demographic and genetic consequences of environmental heterogeneity for biological control. Bull Entom Soc Am 15: 237-240

Levins R (1970) Extinction. In: Gerstenhaber M (ed) Some mathematical questions in biology. American Mathematical Society, Providence, Rhode Island.

Lindenmayer DB, Burgman MA, Akcakaya HR, Lacy RC, Possingham HP (1995) A review of the generic computer programs ALEX, RAMAS/space and VORTEX for modelling the viability of wildlife metapopulations. Ecol Model 82: 161-174

MacArthur RH, Wilson EO (1967) The theory of island biogeography. Princeton University Press, Princeton

Marshall K, Edwards-Jones G (1998) Reintroducing capercaillie (*Tetrao urogallus*) into southern Scotland: identification of minimum viable populations at potential release sites. Biodiversity and Conservation 7: 275-296

May RM (1988) How many species are there on earth ? Science 241: 1441-1449

May RM (1990) How many species? Phil Trans R Soc Lond B 330: 293-304

McCarthy MA, Burgman MA, Ferson S (1995) Sensitivity analysis for models of population viability. Biol Conserv 73: 93-100

Menges ES (1998) Evaluating extinction risks in plant populations. In: Fiedler PL, Kareiva PM (eds) Conservation biology: for the coming decade. Chapman & Hall, New York, pp 49-65

Menges ES (2000) Applications of population viability analyses in plant conservation. Ecological Bulletins 48: 73-84

Moilanen A (1999) Patch occupancy models of metapopulation dynamics: efficient parameter estimation using statistical inference. Ecology 80: 1031-1041

Moilanen A, Smith AT, Hanski I (1998) Long-term dynamics in a metapopulation of the American Pika. Am Nat 152: 530-542

Moloney KA (1993) Determining process through pattern: reality or fantasy? In: Levin SA, Powell TM, Steele JH (eds) Patch dynamics. Springer, Berlin, pp 61-69

Myers N (1981) Conservation needs and opportunities in tropical moist forests. In: Synge H (ed) The biological aspects of rare plant conservation. Wiley, New York, pp 141-154

Neuert, C. Die Dynamik räumlicher Strukturen in naturnahen Buchenwäldern Mitteleuropas. Doctoral thesis, Philipps-Universität Marburg

Neuert C, Rademacher C, Grundmann V, Wissel C, Grimm V (2001) Struktur und Dynamik von Buchenurwäldern: Ergebnisse des regelbasierten Modells BEFORE. Natursch Landschaftspl 33: 173-183

Nisbet RM, Gurney WSC (1982) Modelling Fluctuating Populations. J.Wiley & Sons, Chichester

Noon BR, Lamberson RH, Boyce MS, Irwin LL (1998) Population viability analysis: a primer on its principal technical concepts. In: Johnson NC, Malk AJ, Sexton WT, Szaro RC (eds) Ecological stewartships: a common reference for ecosystem management. Vol. 2. Elsevier, New York, pp 87-134

Noon BR, McKelvey KS (1996) Management of the spotted owl: a case history in conservation biology. Annu Rev Ecol Syst 27: 135-62

Possingham HP, Lindenmayer DB, McCarthy MA (2000) Population viability analysis. In: Levin SA (ed) Encyclopedia of biodiversity. Academic Press, San Diego, pp 831-843.

Primack RB (1993) Essentials of conservation biology. Sinauer, Sunderland

Quammen D D (1996) The song of the Dodo. Hutchinson, London.

Rademacher C, Neuert C, Grundmann V, Wissel C, Grimm V (2001) Was charakterisiert Buchenurwälder? Untersuchungen der Altersstruktur des Kronendachs und der räumlichen Verteilung der Baumriesen in einem Modellwald mit Hilfe des Simulationsmodells BEFORE. Forstwissenschaftliches Centralblatt 120: 288-302

Ranta E, Kaitala V, Lindström J, Helle E (1997) The Moran effect and synchrony in population dynamics. Oikos 78: 136-142

Reich M, Grimm V (1996) Das Metapopulationskonzept in Ökologie und Naturschutz: Eine kritische Bestandsaufnahme. Z Ökologie Naturschutz 5: 123-139

Renshaw E (1991) Modelling biological populations in space and time. Cambridge University Press, Cambridge

Roughgarden J (1998) A primer for ecological theory. Prentice Hall, Upper Saddle River, N.J.

Shaffer ML (1981) Minimum population sizes for species conservation. BioScience 31: 131-134

Simberloff D (1986) The proximate causes of extinction. In: Raup DM, Jablonski D (eds) Patterns and processes in the history of life. Springer, Berlin, pp 259-276

Sjögren-Gulve P, Hanski I (2000) Metapopulation viability analysis using occupancy models. Ecological Bulletins 48: 53-71

Soulé ME (1987) Viable populations for conservation. Cambridge University Press, Cambridge

Starfield AM (1997) A pragmatic approach to modeling for wildlife managment. J Wildlife Manage 61: 261-270

Starfield AM, Smith KA, Bleloch AL (1990) How to model it: problem solving for the computer age. McGraw-Hill Inc., New York St. Louis San Francisco

Stelter C (1997) Persistenz von kleinen Schmetterlingspopulationen in dynamischen Landschaften: ein Populationsdynamik-Modell. Doctoral thesis, Philipps-Universität Marburg

Stelter C, Reich M, Grimm V, Wissel C (1997) Modelling persistence in dynamic landscapes: lesson from a metapopulation of the grasshopper *Bryodema tuberculata*. J Anim Ecol 66: 508-518

Stephan T (1993) Stochastische Modelle zur Extinktion von Populationen. Doctoral-Thesis, Philipps-Universität Marburg

Stephan T, Brendel U, Wissel C (1995) Ein Modell zur Abschätzung des Auslöschungsrisikos von *Alectoris graeca* im Nationalpark Berchtesgaden. Verh GfÖ 24: 161-167

Stephens PA, Frey-Roos F, Arnold W, Sutherland WJ (2002) Model complexity and population persistence. The alpine marmot as a case study. J Anim Ecol 71: 343-361

Storch I (1993) Habitat selection by capercaillie in summer and autumn: is bilberry important? Oecologia 95: 257-265

Storch I (1995) Annual home ranges and spacing patterns of capercaillie in Central Europe. J Wildlife Manage 59: 392-400

Storch I (2001) Capercaillie. BWP Update. The journal of birds of the Western Palearctic (Oxford University Press, Oxford, UK) 3: 1-24

Storch I, Segelbacher G (2000) Genetic correlates of spatial population structure in central European capercaillie and back grouse: a project in progress. Wildlife Biology 6: 239-243

Verboom J, Lankester K, Metz JAJ (1991) Linking local and regional dynamics in stochastic metapopulation models. Biol J Linn Soc 42: 39-55

Verboom J, Metz JAJ, Meelis E (1993) Metapopulation models for impact assessment of fragmentation. In: Vos CS, Opdam P (eds) Landscape ecology of a stressed environment. Chapman and Hall, London, pp 172-191

Vos CC, Verboom J, Opdam PFM, ter Braak CJF (2001) Toward ecologically scaled landscape indices. Am Nat 183: 24-41

Wahlberg N, Moilanen A, Hanski I (1996) Predicting the occurrence of endangered species in fragmented landscapes. Science 273: 1536-1538

Weaver JL, Paquet PC, Ruggiero LF (1996) Resilience and conservation of large carnivores in the Rocky Mountains. Conserv Biol 10: 964-976

Wiegand T, Jeltsch F, Hanski I, Grimm V (2002) Using pattern-oriented modeling for revealing hidden information: a key for reconciling ecological theory and conservation practice. Oikos (in press)

Wiegand T, Moloney KA, Naves J, Knauer F (1999) Finding the missing link between landscape structure and population dynamics: a spatially explicit perspective. Am Nat 154: 605-627

Wiegand T, Naves J, Stephan T, Fernandez A (1998) Assessing the risk of extinction for the brown bear (Ursus arctos) in the Cordillera Cantabrica; Spain. Ecol Monogr 68: 539-570

Wilson EO (1988) The current state of biological diversity. In: Wilson EO, Peter FM (eds) Biodiversity. National Academy Press, Washington, pp 3-18

Wissel C (1989) Theoretische Ökologie - Eine Einführung. Springer, Berlin Heidelberg New York

Wissel C, Stephan T, Zaschke S-H (1994) Modelling extinction and survival of small populations. In: Remmert H (ed) Minimum animal populations (Ecological Studies 106). Springer, Berlin, pp 67-103

Wissel C, Stöcker S (1991) Extinction of populations by random influences. Theor Popul Biol 39: 315-328

With KA (1997) The application of neutral landscape models in conservation biology. Conserv Biol 11: 1069-1080

Index